Porsche 944
Automotive Repair Manual

by Larry Warren, Chaun Muir and John H Haynes

Member of the Guild of Motoring Writers

Models covered:

All Porsche 944 models, including Turbo
1983 through 1989

(80035-5Q10)

Haynes Group Limited
Haynes North America, Inc.
www.haynes.com

© **Haynes North America, Inc. 1987, 1990**

With permission from Haynes Group Limited

A book in the Haynes Automotive Repair Manual Series

ISBN-10: 1-85010-657-6

ISBN-13: 978-1-85010-657-9

Library of Congress Control Number: 89-82763

90-224

Contents

1983 Porsche 944

About this manual

Its purpose

The purpose of this manual is to help you get the best value from your vehicle. It can do so in several ways. It can help you decide what work must be done, even if you choose to have it done by a dealer service department or a repair shop; it provides information and procedures for routine maintenance and servicing; and it offers diagnostic and repair procedures to follow when trouble occurs.

It is hoped that you will use the manual to tackle the work yourself. For many simpler jobs, doing it yourself may be quicker than arranging an appointment to get the vehicle into a shop and making the trips to leave it and pick it up. More importantly, a lot of money can be saved by avoiding the expense the shop must pass on to you to cover its labor and overhead costs. An added benefit is the sense of satisfaction and accomplishment that you feel after having done the job yourself.

Using the manual

The manual is divided into Chapters. Each Chapter is divided into numbered Sections, which are headed in bold type between horizontal lines. Each Section consists of consecutively numbered paragraphs.

At the beginning of each numbered section you will be referred to any illustrations which apply to the procedures in that section. The reference numbers used in illustration captions pinpoint the pertinent Section and the Step within that section. That is, illustration 3.2 means the illustration refers to Section 3 and Step (or paragraph) 2 within that Section.

Procedures, once described in the text, are not normally repeated. When it is necessary to refer to another Chapter, the reference will be given as Chapter and Section number i.e. Chapter 1/16). Cross references given without use of the word ''Chapter'' apply to Sections and/or paragraphs in the same Chapter. For example, ''see Section 8'' means in the same Chapter.

Reference to the left or right side of the vehicle is based on the assumption that one is sitting in the driver's seat, facing forward.

Even though extreme care has been taken during the preparation of this manual, neither the publisher nor the author can accept responsibility for any errors in, or omissions from, the information given.

NOTE

A Note provides information necessary to properly complete a procedure or information which will make the steps to be followed easier to understand.

CAUTION

A Caution indicates a special procedure or special steps which must be taken in the course of completing the procedure in which the **Caution** is found which are necessary to avoid damage to the assembly being worked on.

WARNING

A Warning indicates a special procedure or special steps which must be taken in the course of completing the procedure in which the **Warning** is found which are necessary to avoid injury to the person performing the procedure.

Introduction to the Porsche 944

The Porsche 944, despite its conventional front engine/rear wheel drive layout, incorporates many sophisticated design features.

The fuel-injected engine has an aluminum head and cylinder block. For smoother operation, belt-driven counter-rotating balance shafts mounted in the cylinder block on opposite sides of the crankshaft are used to cancel out vibrations inherent in a four-cylinder engine. Turbocharging is available on later models.

The front-mounted engine and rear-mounted transmission are connected by a central tube. Power is transferred from the engine to either a five-speed manual or three-speed automatic transmission by a one piece driveshaft inside the tube. The rear wheels are driven by driveaxles incorporating constant velocity (CV) joints.

Suspension is independent at all four wheels with coil spring/shock units at the front and torsion bars with shock absorbers at the rear.

The rack and pinion steering gear is mounted in front of the engine and power assist is optional. The brakes are disc on all four wheels, with vacuum assist as standard equipment.

Vehicle Identification Numbers

Modifications are a continuing and unpublicized process in vehicle manufacturing. Since spare parts manuals and lists are compiled on a numerical basis, the individual vehicle numbers are essential to correctly identify the component required.

Vehicle Identification Number (VIN)

This very important identification number is located on a plate attached to the left side windshield post. The VIN also appears on the Vehicle Certificate of Title and Registration. It contains valuable information such as where and when the vehicle was manufactured, the model year and the body style.

Vehicle Identification Label

This label is located in the luggage compartment adjacent to the left tail light under the carpeting. The Vehicle Identification Label contains the VIN, Vehicle Code, Engine and Transmission Code, Paint and Interior Code and Option Codes.

Body code plate

This metal plate is located on the firewall in the engine compartment, adjacent to the battery. Like the VIN, it contains valuable information concerning the production of the vehicle.

Paint number code sticker

The paint number code sticker is found in the engine compartment near the left hood support. This sticker is especially useful for matching the color and type of paint during repair work.

Engine number

The engine number is stamped into the rear of the block, just above the bellhousing, near the base of the oil filler tube.

Transmission identification number

The transmission identification number is stamped into the boss on the top surface of the input shaft housing on manual transmissions and at the bottom surface on automatic transmissions.

Vehicle Emissions Control Information label

The Emissions Control Information label is attached to the front edge of the hood, on the underside (see Chapter 6 for an illustration of the label and its location).

The Vehicle Identification Number (VIN) (arrow) is visible on the driver's side windshield pillar

The vehicle identification label is found in the trunk compartment

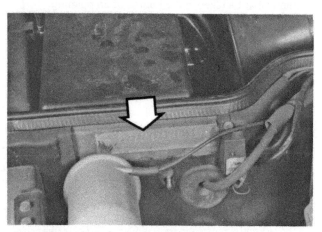

The body code plate (arrow) is located in front of the battery

The paint code sticker (arrow) is located on the left side of the engine compartment

Buying parts

Replacement parts are available from many sources, which generally fall into one of two categories – authorized dealer parts departments and independent retail auto parts stores. Our advice concerning these parts is as follows:

Retail auto parts stores: Good auto parts stores will stock frequently needed components which wear out relatively fast, such as clutch components, exhaust systems, brake parts, tune-up parts, etc. These stores often supply new or reconditioned parts on an exchange basis, which can save a considerable amount of money. Discount auto parts stores are often very good places to buy materials and parts needed for general vehicle maintenance such as oil, grease, filters, spark plugs, belts, touch-up paint, bulbs, etc. They also usually sell tools and general accessories, have con-

venient hours, charge lower prices and can often be found not far from home.

Authorized dealer parts department: This is the best source for parts which are unique to the vehicle and not generally available elsewhere (such as major engine parts, transmission parts, trim pieces, etc.).

Warranty information: If the vehicle is still covered under warranty, be sure that any replacement parts purchased – regardless of the source – do not invalidate the warranty!

To be sure of obtaining the correct parts, have engine and chassis numbers available and, if possible, take the old parts along for positive identification.

Maintenance techniques, tools and working facilities

Maintenance techniques

There are a number of techniques involved in maintenance and repair that will be referred to throughout this manual. Application of these techniques will enable the home mechanic to be more efficient, better organized and capable of performing the various tasks properly, which will ensure that the repair job is thorough and complete.

Fasteners

Fasteners are nuts, bolts, studs and screws used to hold two or more parts together. There are a few things to keep in mind when working with fasteners. Almost all of them use a locking device of some type, either a lockwasher, locknut, locking tab or thread adhesive. All threaded fasteners should be clean and straight, with undamaged threads and undamaged corners on the hex head where the wrench fits. Develop the habit of replacing all damaged nuts and bolts with new ones. Special locknuts with nylon or fiber inserts can only be used

once. If they are removed, they lose their locking ability and must be replaced with new ones.

Rusted nuts and bolts should be treated with a penetrating fluid to ease removal and prevent breakage. Some mechanics use turpentine in a spout-type oil can, which works quite well. After applying the rust penetrant, let it work for a few minutes before trying to loosen the nut or bolt. Badly rusted fasteners may have to be chiseled or sawed off or removed with a special nut breaker, available at tool stores.

If a bolt or stud breaks off in an assembly, it can be drilled and removed with a special tool commonly available for this purpose. Most automotive machine shops can perform this task, as well as other repair procedures, such as the repair of threaded holes that have been stripped out.

Flat washers and lockwashers, when removed from an assembly, should always be replaced exactly as removed. Replace any damaged washers with new ones. Never use a lockwasher on any soft metal surface (such as aluminum), thin sheet metal or plastic.

Fastener sizes

For a number of reasons, automobile manufacturers are making wider and wider use of metric fasteners. Therefore, it is important to be able to tell the difference between standard (sometimes called U.S. or SAE) and metric hardware, since they cannot be interchanged.

All bolts, whether standard or metric, are sized according to diameter, thread pitch and length. For example, a standard 1/2 — 13 x 1 bolt is 1/2 inch in diameter, has 13 threads per inch and is 1 inch long. An M12 — 1.75 x 25 metric bolt is 12 mm in diameter, has a thread pitch of 1.75 mm (the distance between threads) and is 25 mm long. The two bolts are nearly identical, and easily confused, but they are not interchangeable.

In addition to the differences in diameter, thread pitch and length, metric and standard bolts can also be distinguished by examining the bolt heads. To begin with, the distance across the flats on a standard bolt head is measured in inches, while the same dimension on a metric bolt is sized in millimeters (the same is true for nuts). As a result, a standard wrench should not be used on a metric bolt and a metric wrench should not be used on a standard bolt. Also, most standard bolts have slashes radiating out from the center of the head to denote the grade or strength of the bolt, which is an indication of the amount of torque that can be applied to it. The greater the number of slashes, the greater the strength of the bolt. Grades 0 through 5 are commonly used on automobiles. Metric bolts have a property class (grade) number, rather than a slash, molded into their heads to indicate bolt strength. In this case, the higher the number, the stronger the bolt. Property class numbers 8.8, 9.8 and 10.9 are commonly used on automobiles.

Strength markings can also be used to distinguish standard hex nuts from metric hex nuts. Many standard nuts have dots stamped into one side, while metric nuts are marked with a number. The greater the number of dots, or the higher the number, the greater the strength of the nut.

Metric studs are also marked on their ends according to property class (grade). Larger studs are numbered (the same as metric bolts),

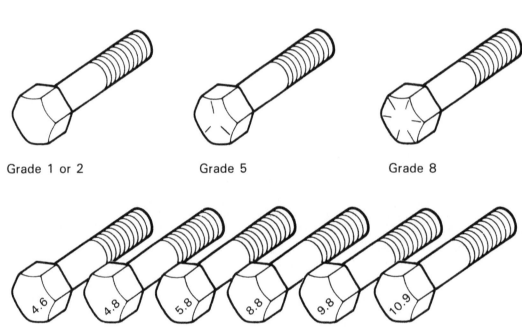

Grade 1 or 2   Grade 5   Grade 8

Bolt strength markings (top — standard/SAE/USS; bottom — metric)

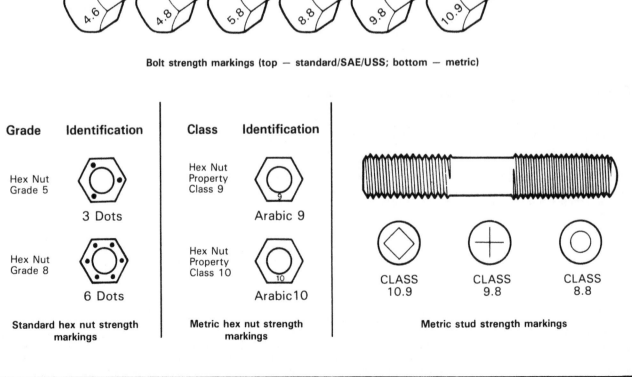

Grade	Identification
Hex Nut Grade 5	3 Dots
Hex Nut Grade 8	6 Dots

Standard hex nut strength markings

Class	Identification
Hex Nut Property Class 9	Arabic 9
Hex Nut Property Class 10	Arabic 10

Metric hex nut strength markings

CLASS 10.9   CLASS 9.8   CLASS 8.8

Metric stud strength markings

while smaller studs carry a geometric code to denote grade.

It should be noted that many fasteners, especially Grades 0 through 2, have no distinguishing marks on them. When such is the case, the only way to determine whether it is standard or metric is to measure the thread pitch or compare it to a known fastener of the same size.

Standard fasteners are often referred to as SAE, as opposed to metric. However, it should be noted that SAE technically refers to a non-metric *fine thread* fastener only. Coarse thread non-metric fasteners are referred to as USS sizes.

Since fasteners of the same size (both standard and metric) may have different strength ratings, be sure to reinstall any bolts, studs or nuts removed from your vehicle in their original locations. Also, when replacing a fastener with a new one, make sure that the new one has a strength rating equal to or greater than the original.

Tightening sequences and procedures

Most threaded fasteners should be tightened to a specific torque value (torque is the twisting force applied to a threaded component such as a nut or bolt). Overtightening the fastener can weaken it and cause it to break, while undertightening can cause it to eventually come loose. Bolts, screws and studs, depending on the material they are made of and their thread diameters, have specific torque values, many of which are noted in the Specifications at the beginning of each Chapter. Be sure to follow the torque recommendations closely. For fasteners not assigned a specific torque, a general torque value chart is presented here as a guide. These torque values are for dry (unlubricated) fasteners threaded into steel or cast iron (not aluminum). As was previously mentioned, the size and grade of a fastener determine the amount of torque that can safely be applied to it. The figures listed here are approximate

Metric thread sizes	Ft-lb	Nm/m
M-6	6 to 9	9 to 12
M-8	14 to 21	19 to 28
M-10	28 to 40	38 to 54
M-12	50 to 71	68 to 96
M-14	80 to 140	109 to 154

Pipe thread sizes	Ft-lb	Nm/m
1/8	5 to 8	7 to 10
1/4	12 to 18	17 to 24
3/8	22 to 33	30 to 44
1/2	25 to 35	34 to 47

U.S. thread sizes	Ft-lb	Nm/m
1/4 — 20	6 to 9	9 to 12
5/16 — 18	12 to 18	17 to 24
5/16 — 24	14 to 20	19 to 27
3/8 — 16	22 to 32	30 to 43
3/8 — 24	27 to 38	37 to 51
7/16 — 14	40 to 55	55 to 74
7/16 — 20	40 to 60	55 to 81
1/2 — 13	55 to 80	75 to 108

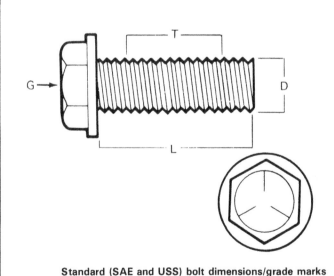

Standard (SAE and USS) bolt dimensions/grade marks

G Grade marks (bolt strength)
L Length (in inches)
T Thread pitch (number of threads per inch)
D Nominal diameter (in inches)

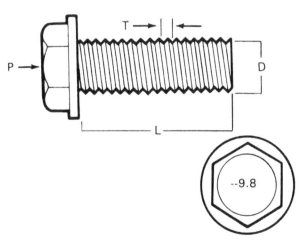

Metric bolt dimensions/grade marks

P Property class (bolt strength)
L Length (in millimeters)
T Thread pitch (distance between threads in millimeters)
D Diameter

for Grade 2 and Grade 3 fasteners. Higher grades can tolerate higher torque values.

Fasteners laid out in a pattern, such as cylinder head bolts, oil pan bolts, differential cover bolts, etc., must be loosened or tightened in sequence to avoid warping the component. This sequence will normally be shown in the appropriate Chapter. If a specific pattern is not given, the following procedures can be used to prevent warping.

Initially, the bolts or nuts should be assembled finger-tight only. Next, they should be tightened one full turn each, in a criss-cross or diagonal pattern. After each one has been tightened one full turn, return to the first one and tighten them all one-half turn, following the same pattern. Finally, tighten each of them one-quarter turn at a time until each fastener has been tightened to the proper torque. To loosen and remove the fasteners, the procedure would be reversed.

Component disassembly

Component disassembly should be done with care and purpose to help ensure that the parts go back together properly. Always keep track of the sequence in which parts are removed. Make note of special characteristics or marks on parts that can be installed more than one way, such as a grooved thrust washer on a shaft. It is a good idea to lay the disassembled parts out on a clean surface in the order that they were removed. It may also be helpful to make sketches or take instant photos of components before removal.

When removing fasteners from a component, keep track of their locations. Sometimes threading a bolt back in a part, or putting the washers and nut back on a stud, can prevent mix-ups later. If nuts and bolts cannot be returned to their original locations, they should be kept in a compartmented box or a series of small boxes. A cupcake or muffin tin is ideal for this purpose, since each cavity can hold the bolts and nuts from a particular area (i.e. oil pan bolts, valve cover bolts, engine mount bolts, etc.). A pan of this type is especially helpful when working on assemblies with very small parts, such as the carburetor, alternator, valve train or interior dash and trim pieces. The cavities can be marked with paint or tape to identify the contents.

Whenever wiring looms, harnesses or connectors are separated, it is a good idea to identify the two halves with numbered pieces of masking tape so they can be easily reconnected.

Gasket sealing surfaces

Throughout any vehicle, gaskets are used to seal the mating surfaces between two parts and keep lubricants, fluids, vacuum or pressure contained in an assembly.

Many times these gaskets are coated with a liquid or paste-type gasket sealing compound before assembly. Age, heat and pressure can sometimes cause the two parts to stick together so tightly that they are very difficult to separate. Often, the assembly can be loosened by striking it with a soft-face hammer near the mating surfaces. A regular hammer can be used if a block of wood is placed between the hammer and the part. Do not hammer on cast parts or parts that could be easily damaged. With any particularly stubborn part, always recheck to make sure that every fastener has been removed.

Avoid using a screwdriver or bar to pry apart an assembly, as they

can easily mar the gasket sealing surfaces of the parts, which must remain smooth. If prying is absolutely necessary, use an old broom handle, but keep in mind that extra clean up will be necessary if the wood splinters.

After the parts are separated, the old gasket must be carefully scraped off and the gasket surfaces cleaned. Stubborn gasket material can be soaked with rust penetrant or treated with a special chemical to soften it so it can be easily scraped off. A scraper can be fashioned from a piece of copper tubing by flattening and sharpening one end. Copper is recommended because it is usually softer than the surfaces to be scraped, which reduces the chance of gouging the part. Some gaskets can be removed with a wire brush, but regardless of the method used, the mating surfaces must be left clean and smooth. If for some reason the gasket surface is gouged, then a gasket sealer thick enough to fill scratches will have to be used during reassembly of the components. For most applications, a non-drying (or semi-drying) gasket sealer should be used.

Hose removal tips

Warning: *If the vehicle is equipped with air conditioning, do not disconnect any of the A/C hoses without first having the system depressurized by a dealer service department or an air conditioning specialist.*

Hose removal precautions closely parallel gasket removal precautions. Avoid scratching or gouging the surface that the hose mates against or the connection may leak. This is especially true for radiator hoses. Because of various chemical reactions, the rubber in hoses can bond itself to the metal spigot that the hose fits over. To remove a hose, first loosen the hose clamps that secure it to the spigot. Then, with slip-joint pliers, grab the hose at the clamp and rotate it around the spigot. Work it back and forth until it is completely free, then pull it off. Silicone or other lubricants will ease removal if they can be applied between the hose and the outside of the spigot. Apply the same lubricant to the inside of the hose and the outside of the spigot to simplify installation.

As a last resort (and if the hose is to be replaced with a new one anyway), the rubber can be slit with a knife and the hose peeled from the spigot. If this must be done, be careful that the metal connection is not damaged.

If a hose clamp is broken or damaged, do not reuse it. Wire-type clamps usually weaken with age, so it is a good idea to replace them with screw-type clamps whenever a hose is removed.

Tools

A selection of good tools is a basic requirement for anyone who plans to maintain and repair his or her own vehicle. For the owner who has few tools, the initial investment might seem high, but when compared to the spiraling costs of professional auto maintenance and repair, it is a wise one.

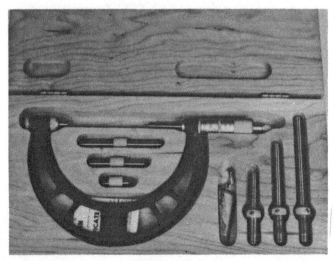

Micrometer set

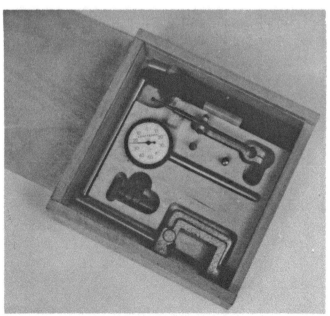

Dial indicator set

Dial caliper

Hand-operated vacuum pump

Timing light

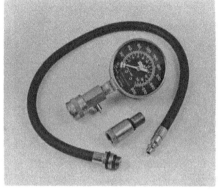

Compression gauge with spark plug
hole adapter

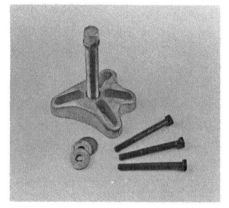

Damper/steering wheel puller

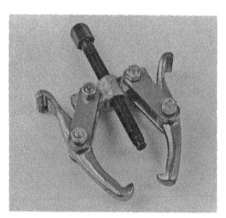

General purpose puller

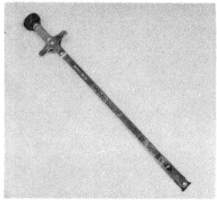

Hydraulic lifter removal tool

Valve spring compressor

Valve spring compressor

Ridge reamer

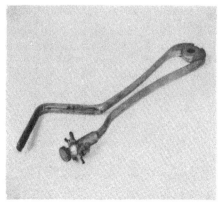

Piston ring groove cleaning tool

Ring removal/installation tool

Ring compressor

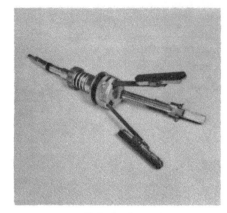

Cylinder hone

Brake hold-down spring tool

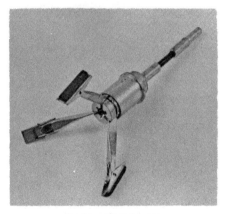

Brake cylinder hone

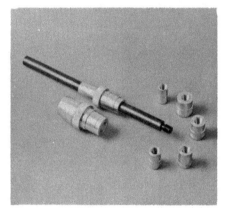

Clutch plate alignment tool

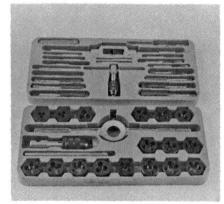

Tap and die set

To help the owner decide which tools are needed to perform the tasks detailed in this manual, the following tool lists are offered: *Maintenance and minor repair*, *Repair/overhaul* and *Special*.

The newcomer to practical mechanics should start off with the maintenance and minor repair tool kit, which is adequate for the simpler jobs performed on a vehicle. Then, as confidence and experience grow, the owner can tackle more difficult tasks, buying additional tools as they are needed. Eventually the basic kit will be expanded into the repair and overhaul tool set. Over a period of time, the experienced do-it-yourselfer will assemble a tool set complete enough for most repair and overhaul procedures and will add tools from the special category when it is felt that the expense is justified by the frequency of use.

Maintenance and minor repair tool kit

The tools in this list should be considered the minimum required for performance of routine maintenance, servicing and minor repair work. We recommend the purchase of combination wrenches (box-end and open-end combined in one wrench). While more expensive than open end wrenches, they offer the advantages of both types of wrench.

Combination wrench set (1/4-inch to 1 inch or 6 mm to 19 mm)
Adjustable wrench, 8 inch
Spark plug wrench with rubber insert
Spark plug gap adjusting tool
Feeler gauge set
Brake bleeder wrench
Standard screwdriver (5/16-inch x 6 inch)
Phillips screwdriver (No. 2 x 6 inch)
Combination pliers — 6 inch
Hacksaw and assortment of blades
Tire pressure gauge
Grease gun
Oil can
Fine emery cloth
Wire brush

Battery post and cable cleaning tool
Oil filter wrench
Funnel (medium size)
Safety goggles
Jackstands (2)
Drain pan

Note: *If basic tune-ups are going to be part of routine maintenance, it will be necessary to purchase a good quality stroboscopic timing light and combination tachometer/dwell meter. Although they are included in the list of special tools, it is mentioned here because they are absolutely necessary for tuning most vehicles properly.*

Repair and overhaul tool set

These tools are essential for anyone who plans to perform major repairs and are in addition to those in the maintenance and minor repair tool kit. Included is a comprehensive set of sockets which, though expensive, are invaluable because of their versatility, especially when various extensions and drives are available. We recommend the 1/2-inch drive over the 3/8-inch drive. Although the larger drive is bulky and more expensive, it has the capacity of accepting a very wide range of large sockets. Ideally, however, the mechanic should have a 3/8-inch drive set and a 1/2-inch drive set.

Socket set(s)
Reversible ratchet
Extension — 10 inch
Universal joint
Torque wrench (same size drive as sockets)
Ball peen hammer — 8 ounce
Soft-face hammer (plastic/rubber)
Standard screwdriver (1/4-inch x 6 inch)
Standard screwdriver (stubby — 5/16-inch)
Phillips screwdriver (No. 3 x 8 inch)
Phillips screwdriver (stubby — No. 2)

Pliers — vise grip
Pliers — lineman's
Pliers — needle nose
Pliers — snap-ring (internal and external)
Cold chisel — 1/2-inch
Scribe
Scraper (made from flattened copper tubing)
Centerpunch
Pin punches (1/16, 1/8, 3/16-inch)
Steel rule/straightedge — 12 inch
Allen wrench set (1/8 to 3/8-inch or 4 mm to 10 mm)
A selection of files
Wire brush (large)
Jackstands (second set)
Jack (scissor or hydraulic type)

Note: *Another tool which is often useful is an electric drill motor with a chuck capacity of 3/8-inch and a set of good quality drill bits.*

Special tools

The tools in this list include those which are not used regularly, are expensive to buy, or which need to be used in accordance with their manufacturer's instructions. Unless these tools will be used frequently, it is not very economical to purchase many of them. A consideration would be to split the cost and use between yourself and a friend or friends. In addition, most of these tools can be obtained from a tool rental shop on a temporary basis.

This list primarily contains only those tools and instruments widely available to the public, and not those special tools produced by the vehicle manufacturer for distribution to dealer service departments. Occasionally, references to the manufacturer's special tools are inluded in the text of this manual. Generally, an alternative method of doing the job without the special tool is offered. However, sometimes there is no alternative to their use. Where this is the case, and the tool cannot be purchased or borrowed, the work should be turned over to the dealer service department or an automotive repair shop.

Valve spring compressor
Piston ring groove cleaning tool
Piston ring compressor
Piston ring installation tool
Cylinder compression gauge
Cylinder ridge reamer
Cylinder surfacing hone
Cylinder bore gauge
Micrometers and/or dial calipers
Hydraulic lifter removal tool
Balljoint separator
Universal-type puller
Impact screwdriver
Dial indicator set
Stroboscopic timing light (inductive pick-up)
Hand operated vacuum/pressure pump
Tachometer/dwell meter
Universal electrical multimeter
Cable hoist
Brake spring removal and installation tools
Floor jack

Buying tools

For the do-it-yourselfer who is just starting to get involved in vehicle maintenance and repair, there are a number of options available when purchasing tools. If maintenance and minor repair is the extent of the work to be done, the purchase of individual tools is satisfactory. If,

on the other hand, extensive work is planned, it would be a good idea to purchase a modest tool set from one of the large retail chain stores. A set can usually be bought at a substantial savings over the individual tool prices, and they often come with a tool box. As additional tools are needed, add-on sets, individual tools and a larger tool box can be purchased to expand the tool selection. Building a tool set gradually allows the cost of the tools to be spread over a longer period of time and gives the mechanic the freedom to choose only those tools that will actually be used.

Tool stores will often be the only source of some of the special tools that are needed, but regardless of where tools are bought, try to avoid cheap ones, especially when buying screwdrivers and sockets, because they won't last very long. The expense involved in replacing cheap tools will eventually be greater than the initial cost of quality tools.

Care and maintenance of tools

Good tools are expensive, so it makes sense to treat them with respect. Keep them clean and in usable condition and store them properly when not in use. Always wipe off any dirt, grease or metal chips before putting them away. Never leave tools lying around in the work area. Upon completion of a job, always check closely under the hood for tools that may have been left there so they won't get lost during a test drive.

Some tools, such as screwdrivers, pliers, wrenches and sockets, can be hung on a panel mounted on the garage or workshop wall, while others should be kept in a tool box or tray. Measuring instruments, gauges, meters, etc. must be carefully stored where they cannot be damaged by weather or impact from other tools.

When tools are used with care and stored properly, they will last a very long time. Even with the best of care, though, tools will wear out if used frequently. When a tool is damaged or worn out, replace it. Subsequent jobs will be safer and more enjoyable if you do.

Working facilities

Not to be overlooked when discussing tools is the workshop. If anything more than routine maintenance is to be carried out, some sort of suitable work area is essential.

It is understood, and appreciated, that many home mechanics do not have a good workshop or garage available, and end up removing an engine or doing major repairs outside. It is recommended, however, that the overhaul or repair be completed under the cover of a roof.

A clean, flat workbench or table of comfortable working height is an absolute necessity. The workbench should be equipped with a vise that has a jaw opening of at least four inches.

As mentioned previously, some clean, dry storage space is also required for tools, as well as the lubricants, fluids, cleaning solvents, etc. which will soon become necessary.

Sometimes waste oil and fluids, drained from the engine or cooling system during normal maintenance or repairs, present a disposal problem. To avoid pouring them on the ground or into a sewage system, pour the used fluids into large containers, seal them with caps and take them to an authorized disposal site or recycling center. Plastic jugs, such as old antifreeze containers, are ideal for this purpose.

Always keep a supply of old newspapers and clean rags available. Old towels are excellent for mopping up spills. Many mechanics use rolls of paper towels for most work because they are readily available and disposable. To help keep the area under the vehicle clean, a large cardboard box can be cut open and flattened to protect the garage or shop floor.

Whenever working over a painted surface, such as when leaning over a fender to service something under the hood, always cover it with an old blanket or bedspread to protect the finish. Vinyl covered pads, made especially for this purpose, are available at auto parts stores.

Booster battery (jump) starting

Certain precautions must be observed when using a booster battery to jump start a vehicle.

a) The booster battery must the same voltage (12 volts) as the dead one in the vehicle.
b) If the booster battery is installed in another vehicle, the two vehicles must not be touching one another.
c) The lights, heater, air conditioner and all other electrical components must be turned off.
d) The transmission must be in Neutral (manual) or Park (automatic).
e) The ignition switch must be turned to the Off position before the booster cables are hooked up.
f) Batteries contain sulfuric acid, which is poisonous and corrosive. Protect your eyes with safety glasses or goggles. Avoid spilling acid on your skin, clothing or the vehicle. Should you accidentally get acid on your skin or in your eyes, remove any contaminated clothing immediately and flush the exposed area with water for at least 15 minutes, then get immediate medical attention. If possible, continue to apply water to the affected area with a sponge or damp cloth enroute to the medical office.
g) The vent caps must be removed and the vent holes must be covered by a cloth.
h) There must be no smoking or open flames in the vicinity of the battery. The hydrogen gas produced by the battery will explode if it comes into contact with a flame or spark. If the engine in the vehicle with the booster battery is not running, start it and let it run for a few minutes. Once it is warmed up, turn it off.

Connect the jumper cables exactly as shown (see illustration). Connect the red (positive) jumper cable to the *Positive* (+) terminals of each battery.

Connect one end of the black jumper cable to the *negative* (–) terminal of the booster battery. The other end of this cable should be connected to a good ground on the vehicle to be started, such as a bolt or bracket on the engine block. Make sure that the cable does not come into contact with any moving parts of the engine.

Start the engine in the vehicle with the booster battery and, during jump starting, run it at about 2000 rpm. Start the engine with the dead battery. Let it idle, then disconnect the jumper cables in the reverse order of connection.

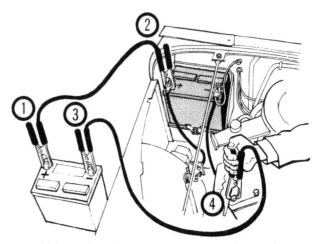

Make the booster battery cable connections in the numerical order shown (note that the negative cable of the booster battery *is not* attached to the negative terminal of the dead battery)

Jacking and towing

Jacking

The jack supplied with the vehicle should only be used for raising the vehicle when changing a tire or placing jackstands under the frame. **Warning:** *Never work under the vehicle or start the engine while this jack is being used as the only means of support.*

The vehicle should be on level ground with the wheels blocked and the transmission in Park (automatic) or Reverse (manual). If the wheel is being replaced, loosen the wheel nuts one-half turn and leave them in place until the wheel is raised off the ground. Refer to Chapter 1 for information related to removing and installing the tire.

Place the jack under the side of the vehicle in the indicated position and raise it until the jack head hole fits over the rocker flange jack locator pin. Operate the jack with a slow, smooth motion until the wheel is raised off the ground.

Lower the vehicle, remove the jack and tighten the nuts (if loosened or removed) in a criss-cross sequence by turning the wrench clockwise.

Towing

Vehicles with an automatic transmission can be towed with all four wheels on the ground, provided that speeds do not exceed 30 mph and the distance is not over 30 miles, otherwise transmission damage can result. Vehicles with a manual transmission can be towed at legal highway speeds for any distance. If the vehicle has a damaged transmission, tow it only with the rear wheels off the ground.

Towing equipment specifically designed for this purpose should be used and should be attached to the main structural members of the vehicle, not the bumper or brackets.

Safety is a major consideration when towing and all applicable state and local laws must be obeyed. A safety chain system must be used for all towing.

While towing, the parking brake should be released and the transmission must be in Neutral. The steering must be unlocked (ignition switch in the Off position). Remember that power steering and power brakes will not work with the engine off.

Lifting points for workshop hoist or floor jack (white arrows), and jackport for car jack (black arrows)

Automotive chemicals and lubricants

A number of automotive chemicals and lubricants are available for use during vehicle maintenance and repair. They include a wide variety of products ranging from cleaning solvents and degreasers to lubricants and protective sprays for rubber, plastic and vinyl.

Cleaners

Carburetor cleaner and choke cleaner is a strong solvent for gum, varnish and carbon. Most carburetor cleaners leave a dry-type lubricant film which will not harden or gum up. Because of this film it is not recommended for use on electrical components.

Brake system cleaner is used to remove grease and brake fluid from the brake system where clean surfaces are absolutely necessary. It leaves no residue and often eliminates brake squeal caused by contaminants.

Electrical cleaner removes oxidation, corrosion and carbon deposits from electrical contacts, restoring full current flow. It can also be used to clean spark plugs, carburetor jets, voltage regulators and other parts where an oil-free surface is desired.

Demoisturants remove water and moisture from electrical components such as alternators, voltage regulators, electrical connectors and fuse blocks. It is non-conductive, non-corrosive and non-flammable.

Degreasers are heavy-duty solvents used to remove grease from the outside of the engine and from chassis components. They can be sprayed or brushed on, and, depending on the type, are rinsed off either with water or solvent.

Lubricants

Motor oil is the lubricant formulated for use in engines. It normally contains a wide variety of additives to prevent corrosion and reduce foaming and wear. Motor oil comes in various weights (viscosity ratings) from 5 to 80. The recommended weight of the oil depends on the season, temperature and the demands on the engine. Light oil is used in cold climates and under light load conditions. Heavy oil is used in hot climates and where high loads are encountered. Multi-viscosity oils are designed to have characteristics of both light and heavy oils and are available in a number of weights from 5W-20 to 20W-50.

Gear oil is designed to be used in differentials, manual transaxles and other areas where high-temperature lubrication is required.

Chassis and wheel bearing grease is a heavy grease used where increased loads and friction are encountered, such as for wheel bearings, balljoints, tie rod ends and universal joints.

High temperature wheel bearing grease is designed to withstand the extreme temperatures encountered by wheel bearings in disc brake equipped vehicles. It usually contains molybdenun disulfide (moly), which is a dry-type lubricant.

White grease is a heavy grease for metal to metal applications where water is a problem. White grease stays soft under both low and high temperatures (usually from $-100\,^\circ F$ to $+190\,^\circ F$), and will not wash off or dilute in the presence of water.

Assembly lube is a special extreme pressure lubricant, usually containing moly, used to lubricate high-load parts such as main and rod bearings and cam lobes for initial start-up of a new engine. The assembly lube lubricates the parts without being squeezed out or washed away until the engine oiling system begins to function.

Silicone lubricants are used to protect rubber, plastic, vinyl and nylon parts.

Graphite lubricants are used where oils cannot be used due to contamination problems, such as in locks. The dry graphite will lubricate metal parts while remaining uncontaminated by dirt, water, oil or acids. It is electrically conductive and will not foul electrical contacts in locks such as the ignition switch.

Moly penetrants loosen and lubricate frozen, rusted and corroded fasteners and prevent future rusting or freezing.

Heat-sink grease is a special electrically non-conductive grease that is used for mounting HEI ignition modules where it is essential that heat be transferred away from the module.

Sealants

RTV sealant is one of the most widely used gasket compounds. Made from silicone, RTV is air curing, it seals, bonds, waterproofs, fills surface irregularities, remains flexible, doesn't shrink, is relatively easy to remove, and is used as a supplementary sealer with almost all low and medium temperature gaskets.

Anaerobic sealant is much like RTV in that it can be used either to seal gaskets or to form gaskets by itself. It remains flexible, is solvent resistant and fills surface imperfections. The difference between an anaerobic sealant and an RTV-type sealant is in the curing. RTV cures when exposed to air, while an anaerobic sealant cures only in the absence of air. This means that an anaerobic sealant cures only after the assembly of parts, sealing them together.

Thread and pipe sealant is used for sealing hydraulic and pneumatic fittings and vacuum lines. It is usually made from a teflon compound, and comes in a spray, a paint-on liquid and as a wrap-around tape.

Chemicals

Anti-seize compound prevents seizing, galling, cold welding, rust and corrosion in fasteners. High temperature anti-seize, usually made with copper and graphite lubricants, is used for exhaust system and manifold bolts.

Anaerobic locking compounds are used to keep fasteners from vibrating or working loose, and cure only after installation, in the absence of air. Medium strength locking compound is used for small nuts, bolts and screws that you expect to be removing later. High strength locking compound is for large nuts, bolts and studs which you don't intend to be removing on a regular basis.

Oil additives range from viscosity index improvers to chemical treatments that claim to reduce internal engine friction. It should be noted that most oil manufacturers caution against using additives with their oils.

Gas additives perform several functions, depending on their chemical makeup. They usually contain solvents that help dissolve gum and varnish that build up on carburetor and intake parts. They also serve to break down carbon deposits that form on the inside surfaces of the combustion chambers. Some additives contain upper cylinder lubricants for valves and piston rings, and others chemicals to remove condensation from the gas tank.

Miscellaneous

Brake fluid is specially formulated hydraulic fluid that can withstand the heat and pressure encountered in brake systems. Care must be taken that this fluid does not come in contact with painted surfaces or plastics. An opened container should always be resealed to prevent contamination by water or dirt.

Weatherstrip adhesive is used to bond weatherstripping around doors, windows and trunk lids. It is sometimes used to attach trim pieces.

Undercoating is a petroleum-based tar-like substance that is designed to protect metal surfaces on the underside of the vehicle from corrosion. It also acts as a sound-deadening agent by insulating the bottom of the vehicle.

Waxes and polishes are used to help protect painted and plated surfaces from the weather. Different types of paint may require the use of different types of wax and polish. Some polishes utilize a chemical or abrasive cleaner to help remove the top layer of oxidized (dull) paint on older vehicles. In recent years many non-wax polishes that contain a wide variety of chemicals such as polymers and silicones have been introduced. These non-wax polishes are usually easier to apply and last longer than conventional waxes and polishes.

Safety first!

Regardless of how enthusiastic you may be about getting on with the job at hand, take the time to ensure that your safety is not jeopardized. A moment's lack of attention can result in an accident, as can failure to observe certain simple safety precautions. The possibility of an accident will always exist, and the following points should not be considered a comprehensive list of all dangers. Rather, they are intended to make you aware of the risks and to encourage a safety conscious approach to all work you carry out on your vehicle.

Essential DOs and DON'Ts

DON'T rely on a jack when working under the vehicle. Always use approved jackstands to support the weight of the vehicle and place them under the recommended lift or support points.

DON'T attempt to loosen extremely tight fasteners (i.e. wheel lug nuts) while the vehicle is on a jack — it may fall.

DON'T start the engine without first making sure that the transmission is in Neutral (or Park where applicable) and the parking brake is set.

DON'T remove the radiator cap from a hot cooling system — let it cool or cover it with a cloth and release the pressure gradually.

DON'T attempt to drain the engine oil until you are sure it has cooled to the point that it will not burn you.

DON'T touch any part of the engine or exhaust system until it has cooled sufficiently to avoid burns.

DON'T siphon toxic liquids such as gasoline, antifreeze and brake fluid by mouth, or allow them to remain on your skin.

DON'T inhale brake lining dust — it is potentially hazardous (see *Asbestos* below)

DON'T allow spilled oil or grease to remain on the floor — wipe it up before someone slips on it.

DON'T use loose fitting wrenches or other tools which may slip and cause injury.

DON'T push on wrenches when loosening or tightening nuts or bolts. Always try to pull the wrench toward you. If the situation calls for pushing the wrench away, push with an open hand to avoid scraped knuckles if the wrench should slip.

DON'T attempt to lift a heavy component alone — get someone to help you.

DON'T rush or take unsafe shortcuts to finish a job.

DON'T allow children or animals in or around the vehicle while you are working on it.

DO wear eye protection when using power tools such as a drill, sander, bench grinder, etc. and when working under a vehicle.

DO keep loose clothing and long hair well out of the way of moving parts.

DO make sure that any hoist used has a safe working load rating adequate for the job.

DO get someone to check on you periodically when working alone on a vehicle.

DO carry out work in a logical sequence and make sure that everything is correctly assembled and tightened.

DO keep chemicals and fluids tightly capped and out of the reach of children and pets.

DO remember that your vehicle's safety affects that of yourself and others. If in doubt on any point, get professional advice.

Asbestos

Certain friction, insulating, sealing, and other products — such as brake linings, brake bands, clutch linings, torque converters, gaskets, etc. — contain asbestos. *Extreme care must be taken to avoid inhalation of dust from such products since it is hazardous to health*. If in doubt, assume that they *do* contain asbestos.

Fire

Remember at all times that gasoline is highly flammable. Never smoke or have any kind of open flame around when working on a vehicle. But the risk does not end there. A spark caused by an electrical short circuit, by two metal surfaces contacting each other, or even by static electricity built up in your body under certain conditions, can ignite gasoline vapors, which in a confined space are highly explosive. Do not, under any circumstances, use gasoline for cleaning parts. Use an approved safety solvent.

Always disconnect the battery ground (–) cable *at the battery* before working on any part of the fuel system or electrical system. Never risk spilling fuel on a hot engine or exhaust component.

It is strongly recommended that a fire extinguisher suitable for use on fuel and electrical fires be kept handy in the garage or workshop at all times. Never try to extinguish a fuel or electrical fire with water.

Fumes

Certain fumes are highly toxic and can quickly cause unconsciousness and even death if inhaled to any extent. Gasoline vapor falls into this category, as do the vapors from some cleaning solvents. Any draining or pouring of such volatile fluids should be done in a well ventilated area.

When using cleaning fluids and solvents, read the instructions on the container carefully. Never use materials from unmarked containers.

Never run the engine in an enclosed space, such as a garage. Exhaust fumes contain carbon monoxide, which is extremely poisonous. If you need to run the engine, always do so in the open air, or at least have the rear of the vehicle outside the work area.

If you are fortunate enough to have the use of an inspection pit, never drain or pour gasoline and never run the engine while the vehicle is over the pit. The fumes, being heavier than air, will concentrate in the pit with possibly lethal results.

The battery

Never create a spark or allow a bare light bulb near a battery. They normally give off a certain amount of hydrogen gas, which is highly explosive.

Always disconnect the battery ground (–) cable *at the battery* before working on the fuel or electrical systems.

If possible, loosen the filler caps or cover when charging the battery from an external source (this does not apply to sealed or maintenance-free batteries). Do not charge at an excessive rate or the battery may burst.

Take care when adding water to a non maintenance-free battery and when carrying a battery. The electrolyte, even when diluted, is very corrosive and should not be allowed to contact clothing or skin.

Always wear eye protection when cleaning the battery to prevent the caustic deposits from entering your eyes.

Household current

When using an electric power tool, inspection light, etc., which operates on household current, always make sure that the tool is correctly connected to its plug and that, where necessary, it is properly grounded. Do not use such items in damp conditions and, again, do not create a spark or apply excessive heat in the vicinity of fuel or fuel vapor.

Secondary ignition system voltage

A severe electric shock can result from touching certain parts of the ignition system (such as the spark plug wires) when the engine is running or being cranked, particularly if components are damp or the insulation is defective. In the case of an electronic ignition system, the secondary system voltage is much higher and could prove fatal.

Conversion factors

Length (distance)
Inches (in)	X	25.4	= Millimetres (mm)	X	0.0394	= Inches (in)
Feet (ft)	X	0.305	= Metres (m)	X	3.281	= Feet (ft)
Miles	X	1.609	= Kilometres (km)	X	0.621	= Miles

Volume (capacity)
Cubic inches (cu in; in³)	X	16.387	= Cubic centimetres (cc; cm³)	X	0.061	= Cubic inches (cu in; in³)
Imperial pints (Imp pt)	X	0.568	= Litres (l)	X	1.76	= Imperial pints (Imp pt)
Imperial quarts (Imp qt)	X	1.137	= Litres (l)	X	0.88	= Imperial quarts (Imp qt)
Imperial quarts (Imp qt)	X	1.201	= US quarts (US qt)	X	0.833	= Imperial quarts (Imp qt)
US quarts (US qt)	X	0.946	= Litres (l)	X	1.057	= US quarts (US qt)
Imperial gallons (Imp gal)	X	4.546	= Litres (l)	X	0.22	= Imperial gallons (Imp gal)
Imperial gallons (Imp gal)	X	1.201	= US gallons (US gal)	X	0.833	= Imperial gallons (Imp gal)
US gallons (US gal)	X	3.785	= Litres (l)	X	0.264	= US gallons (US gal)

Mass (weight)
Ounces (oz)	X	28.35	= Grams (g)	X	0.035	= Ounces (oz)
Pounds (lb)	X	0.454	= Kilograms (kg)	X	2.205	= Pounds (lb)

Force
Ounces-force (ozf; oz)	X	0.278	= Newtons (N)	X	3.6	= Ounces-force (ozf; oz)
Pounds-force (lbf; lb)	X	4.448	= Newtons (N)	X	0.225	= Pounds-force (lbf; lb)
Newtons (N)	X	0.1	= Kilograms-force (kgf; kg)	X	9.81	= Newtons (N)

Pressure
Pounds-force per square inch (psi; lbf/in²; lb/in²)	X	0.070	= Kilograms-force per square centimetre (kgf/cm²; kg/cm²)	X	14.223	= Pounds-force per square inch (psi; lbf/in²; lb/in²)
Pounds-force per square inch (psi; lbf/in²; lb/in²)	X	0.068	= Atmospheres (atm)	X	14.696	= Pounds-force per square inch (psi; lbf/in²; lb/in²)
Pounds-force per square inch (psi; lbf/in²; lb/in²)	X	0.069	= Bars	X	14.5	= Pounds-force per square inch (psi; lbf/in²; lb/in²)
Pounds-force per square inch (psi; lbf/in²; lb/in²)	X	6.895	= Kilopascals (kPa)	X	0.145	= Pounds-force per square inch (psi; lbf/in²; lb/in²)
Kilopascals (kPa)	X	0.01	= Kilograms-force per square centimetre (kgf/cm²; kg/cm²)	X	98.1	= Kilopascals (kPa)
Millibar (mbar)	X	100	= Pascals (Pa)	X	0.01	= Millibar (mbar)
Millibar (mbar)	X	0.0145	= Pounds-force per square inch (psi; lbf/in²; lb/in²)	X	68.947	= Millibar (mbar)
Millibar (mbar)	X	0.75	= Millimetres of mercury (mmHg)	X	1.333	= Millibar (mbar)
Millibar (mbar)	X	0.401	= Inches of water (inH₂O)	X	2.491	= Millibar (mbar)
Millimetres of mercury (mmHg)	X	0.535	= Inches of water (inH₂O)	X	1.868	= Millimetres of mercury (mmHg)
Inches of water (inH₂O)	X	0.036	= Pounds-force per square inch (psi; lbf/in²; lb/in²)	X	27.68	= Inches of water (inH₂O)

Torque (moment of force)
Pounds-force inches (lbf in; lb in)	X	1.152	= Kilograms-force centimetre (kgf cm; kg cm)	X	0.868	= Pounds-force inches (lbf in; lb in)
Pounds-force inches (lbf in; lb in)	X	0.113	= Newton metres (Nm)	X	8.85	= Pounds-force inches (lbf in; lb in)
Pounds-force inches (lbf in; lb in)	X	0.083	= Pounds-force feet (lbf ft; lb ft)	X	12	= Pounds-force inches (lbf in; lb in)
Pounds-force feet (lbf ft; lb ft)	X	0.138	= Kilograms-force metres (kgf m; kg m)	X	7.233	= Pounds-force feet (lbf ft; lb ft)
Pounds-force feet (lbf ft; lb ft)	X	1.356	= Newton metres (Nm)	X	0.738	= Pounds-force feet (lbf ft; lb ft)
Newton metres (Nm)	X	0.102	= Kilograms-force metres (kgf m; kg m)	X	9.804	= Newton metres (Nm)

Power
Horsepower (hp)	X	745.7	= Watts (W)	X	0.0013	= Horsepower (hp)

Velocity (speed)
Miles per hour (miles/hr; mph)	X	1.609	= Kilometres per hour (km/hr; kph)	X	0.621	= Miles per hour (miles/hr; mph)

Fuel consumption*
Miles per gallon, Imperial (mpg)	X	0.354	= Kilometres per litre (km/l)	X	2.825	= Miles per gallon, Imperial (mpg)
Miles per gallon, US (mpg)	X	0.425	= Kilometres per litre (km/l)	X	2.352	= Miles per gallon, US (mpg)

Temperature

Degrees Fahrenheit = (°C x 1.8) + 32

Degrees Celsius (Degrees Centigrade; °C) = (°F - 32) x 0.56

*It is common practice to convert from miles per gallon (mpg) to litres/100 kilometres (l/100km), where mpg (Imperial) x l/100 km = 282 and mpg (US) x l/100 km = 235

Troubleshooting

Contents

This section provides an easy reference guide to the more common problems which may occur during the operation of your vehicle. These problems and their possible causes are grouped under headings denoting various components or systems, such as Engine, Cooling system, etc. They also refer you to the Chapter and/or Section which deals with the problem.

Remember that successful troubleshooting is not a mysterious black art practiced only by professional mechanics. It is simply the result of the right knowledge combined with an intelligent, systematic approach to the problem. Always work by a process of elimination, starting with the simplest solution and working through to the most complex — and never overlook the obvious. Anyone can run the gas tank dry or leave the lights on overnight, so don't assume that you are exempt from such oversights.

Finally, always establish a clear idea of why a problem has occurred and take steps to ensure that it doesn't happen again. If the electrical system fails because of a poor connection, check all other connections in the system to make sure that they don't fail as well. If a particular fuse continues to blow, find out why — don't just replace one fuse after another. Remember, failure of a small component can often be indicative of potential failure or incorrect functioning of a more important component or system.

Engine

1 Engine will not rotate when attempting to start

1 Battery terminal connections loose or corroded (Chapter 1).
2 Battery discharged or faulty (Chapter 1).
3 Automatic transmission not completely engaged in Park.
4 Broken, loose or disconnected wiring in the starting circuit (Chapters 5 and 12).
5 Starter motor pinion jammed in flywheel ring gear (Chapter 5).
6 Starter solenoid faulty (Chapter 5).
7 Starter motor faulty (Chapter 5).
8 Ignition switch faulty (Chapter 12).
9 Starter pinion or flywheel teeth worn or broken (Chapter 5).

2 Engine rotates but will not start

1 Fuel tank empty.
2 Battery discharged (engine rotates slowly) (Chapter 5).
3 Battery terminal connections loose or corroded (Chapter 1).
4 Leaking fuel injector(s), faulty Digital Engine Electronics (DEE) components, fuel pump, pressure regulator, etc. (Chapter 4).
5 Fuel not reaching fuel injection system (Chapter 4).
6 Ignition components damp or damaged (Chapter 5).
7 Worn, faulty or incorrectly gapped spark plugs (Chapter 1).
8 Broken, loose or disconnected wiring in the starting circuit (Chapter 5).
9 Broken, loose or disconnected wires at the ignition coil or faulty coil (Chapter 5).

3 Engine hard to start when cold

1 Battery discharged or low (Chapter 1).
2 Fuel system malfunctioning (Chapter 4).
3 Injector(s) leaking (Chapter 4).
4 Digital Engine Electronics (DEE) system fault (Chapter 4).

4 Engine hard to start when hot

1 Air filter clogged (Chapter 1).
2 Fuel not reaching the fuel injection system (Chapter 4).
3 Corroded battery connections, especially ground (Chapter 1).

5 Starter motor noisy or excessively rough in engagement

1 Pinion or flywheel gear teeth worn or broken (Chapter 5).
2 Starter motor mounting bolts loose or missing (Chapter 5).

6 Engine starts but stops immediately

1 Loose or faulty electrical connections at distributor, coil or alternator (Chapter 5).
2 Insufficient fuel reaching the fuel injector(s) (Chapters 1 and 4).

7 Oil puddle under engine

1 Oil pan gasket and/or oil pan drain bolt seal leaking (Chapter 2).
2 Oil pressure sending unit leaking (Chapter 2).
3 Engine oil seals leaking (Chapter 2).
4 Leaking balance shaft seals (Chapter 2).

8 Engine lopes while idling or idles erratically

1 Vacuum leakage (Chapter 4).
2 Air filter clogged (Chapter 1).
3 Fuel pump not delivering sufficient fuel to the fuel injection system (Chapter 4).
4 Leaking head gasket (Chapter 2).
5 Camshaft lobes worn (Chapter 2).

9 Engine misses at idle speed

1 Spark plugs worn or not gapped properly (Chapter 1).
2 Faulty spark plug wires (Chapter 1).
3 Vacuum leaks (Chapter 1).
4 Uneven or low compression (Chapter 2).
5 Digital Engine Electronics (DEE) system fault (Chapter 4).

10 Engine misses throughout driving speed range

1 Fuel filter clogged and/or impurities in the fuel system (Chapter 1).
2 Low fuel output at the injector (Chapter 4).
3 Faulty or incorrectly gapped spark plugs (Chapter 1).
4 Digital Engine Electronics (DEE) system fault (Chapter 4).
5 Cracked distributor cap, disconnected distributor wires or damaged distributor components (Chapter 1).
6 Leaking spark plug wires (Chapter 1).
7 Low or uneven cylinder compression pressures (Chapter 1).
8 Weak or faulty ignition system (Chapter 5).
9 Vacuum leak in fuel injection system, intake manifold or vacuum hoses (Chapter 4).
10 Digital Engine Electronics (DEE) system fault (Chapter 4).

11 Engine stumbles on acceleration

1 Spark plugs fouled (Chapter 1).
2 Fuel injection system needs adjustment or repair (Chapter 4).
3 Fuel filter clogged (Chapters 1 and 4).
4 Digital Engine Electronics (DEE) system fault (Chapter 4).

12 Engine surges while holding accelerator steady

1 Digital Engine Electronics (DEE) system fault (Chapter 4).
2 Fuel pump faulty (Chapter 4).
3 Loose fuel injector harness connections (Chapters 4 and 6).
4 Defective control unit (Chapter 6).

13 Engine stalls

1 Idle speed incorrect (Chapter 1).
2 Fuel filter clogged and/or water and impurities in the fuel system (Chapter 1).
3 Distributor cap and rotor damp or damaged (Chapter 5).
5 Faulty or incorrectly gapped spark plugs (Chapter 1).
6 Faulty spark plug wires (Chapter 1).
7 Vacuum leak in the fuel injection system, intake manifold or vacuum hoses (Chapter 4).

14 Engine lacks power

1 Digital Engine Electronics (DEE) system fault (Chapter 4).
2 Worn rotor, distributor cap or wires (Chapters 1 and 5).
3 Faulty or incorrectly gapped spark plugs (Chapter 1).
4 Fuel injection system out of adjustment or excessively worn (Chapter 4).
5 Faulty coil (Chapter 5).
6 Brakes binding (Chapter 1).
7 Automatic transmission fluid level incorrect (Chapter 1).
8 Clutch slipping (Chapter 8).
9 Fuel filter clogged and/or impurities in the fuel system (Chapter 1).
10 Low or uneven cylinder compression pressures (Chapter 1).

15 Engine backfires

1 Faulty secondary ignition system (cracked spark plug insulator, faulty plug wires, distributor cap and/or rotor) (Chapters 1 and 5).
2 Fuel injection system in need of adjustment or worn excessively (Chapter 4).
3 Vacuum leak at fuel injector(s), intake manifold or vacuum hoses (Chapter 4).
4 Digital Engine Electronics (DEE) system fault (Chapter 4).

16 Pinging or knocking engine sounds during acceleration or uphill

1 Incorrect grade of fuel.
2 Digital Engine Electronics (DEE) system fault (Chapter 4).
3 Fuel injection system in need of adjustment (Chapter 4).
4 Improper or damaged spark plugs or wires (Chapter 1).
5 Worn or damaged distributor components (Chapter 5).
6 Vacuum leak (Chapter 4).

17 Engine runs with oil pressure light on

1 Low oil level (Chapter1).
2 Digital Engine Electronics (DEE) system fault (Chapter 4).
3 Short in wiring circuit (Chapter 12).
4 Faulty oil pressure sender (Chapter 2).
5 Worn engine bearings and/or oil pump (Chapter 2).

18 Engine diesels (continues to run) after switching off

1 Digital Engine Electronics (DEE) system fault (Chapter 4).
2 Excessive engine operating temperature (Chapter 3).

Engine electrical system

19 Battery will not hold a charge

1 Alternator drivebelt defective or not adjusted properly (Chapter 1).
2 Electrolyte level low (Chapter 1).

3 Battery terminals loose or corroded (Chapter 1).
4 Alternator not charging properly (Chapter 5).
5 Loose, broken or faulty wiring in the charging circuit (Chapter 5).
6 Short in vehicle wiring (Chapters 5 and 12).
7 Internally defective battery (Chapters 1 and 5).

20 Voltage warning light fails to go out

1 Faulty alternator or charging circuit (Chapter 5).
2 Alternator drivebelt defective or out of adjustment (Chapter 1).
3 Alternator voltage regulator inoperative (Chapter 5).

21 Voltage warning light fails to come on when key is turned on

1 Warning light bulb defective (Chapter 12).
2 Fault in the circuit, dash wiring or bulb holder (Chapter 12).

Fuel system

22 Excessive fuel consumption

1 Dirty or clogged air filter element (Chapter 1).
2 Digital Engine Electronics (DEE) system fault (Chapter 4).
3 Fuel injection internal parts excessively worn or damaged (Chapter 4).
4 Low tire pressure or incorrect tire size (Chapter 1).

23 Fuel leakage and/or fuel odor

1 Leak in a fuel feed or vent line (Chapter 4).
2 Tank overfilled.
3 Evaporative canister filter clogged (Chapters 1 and 6).
4 Fuel injector internal parts excessively worn (Chapter 4).

Cooling system

24 Overheating

1 Insufficient coolant in system (Chapter 1).
2 Water pump defective (Chapter 1).
3 Radiator core blocked or grille restricted (Chapter 3).
4 Thermostat faulty (Chapter 3).
5 Electric coolant fan motor or wiring faulty (Chapter 3).
6 Coolant cap not maintaining proper pressure (Chapter 3).
7 Digital Engine Electronics (DEE) system fault (Chapter 4).

25 Overcooling

Faulty thermostat (Chapter 3).

26 External coolant leakage

1 Deteriorated/damaged hoses; loose clamps (Chapters 1 and 3).
2 Water pump seal defective (Chapters 1 and 3).
3 Leakage from radiator core, reservoir or header tank (Chapter 3).
4 Engine drain or bleed screws leaking (Chapter 2).

27 Internal coolant leakage

1 Leaking cylinder head gasket (Chapter 2).
2 Cracked cylinder bore or cylinder head (Chapter 2).
3 Damaged or faulty oil cooler.

28 Coolant loss

1   Too much coolant in system (Chapter 1).
2   Coolant boiling away because of overheating (Chapter 3).
3   Internal or external leakage (Chapter 3).
4   Faulty coolant cap (Chapter 3).

29 Poor coolant circulation

1   Inoperative water pump (Chapter 3).
2   Restriction in cooling system (Chapters 1 and 3).
3   Water pump defective.
4   Thermostat sticking (Chapter 3).

Clutch

30 Pedal travels to floor — no pressure or very little resistance

1   Master or slave cylinder faulty (Chapter 8).
2   Hose/pipe burst or leaking (Chapter 8).
3   Connections leaking (Chapter 8).
4   No fluid in reservoir (Chapter 8).
5   If fluid level in reservoir rises as pedal is depressed, master cylinder center valve seal is faulty (Chapter 8).
7   Broken release bearing or lever (Chapter 8).

31 Fluid on slave cylinder

Slave cylinder internal fault (Chapter 8).

32 Pedal feels 'spongy' when depressed

Air in system (Chapter 8).

33 Unable to select gears

1   Faulty transmission (Chapter 7).
2   Faulty clutch disc (Chapter 8).
3   Release lever and bearing not assembled properly (Chapter 8).
4   Faulty pressure plate (Chapter 8).
5   Pressure plate-to-flywheel bolts loose (Chapter 8).

34 Clutch slips (engine speed increases with no increase in vehicle speed)

1   Clutch plate worn (Chapter 8).
2   Clutch plate not seated. It may take 30 or 40 normal starts for a new one to seat.
3   Warped pressure plate or flywheel (Chapter 8).
4   Weak diaphragm spring (Chapter 8).
5   Clutch plate overheated. Allow to cool.

35 Grabbing (chattering) as clutch is engaged

1   Oil on clutch plate lining, burned or glazed facings (Chapter 8).
2   Worn or loose engine or transmission mounts (Chapters 2 and 7).
3   Worn splines on clutch plate hub (Chapter 8).
4   Warped pressure plate or flywheel (Chapter 8).

36 Noise in clutch area

1   Fork shaft improperly installed (Chapter 8).
2   Faulty bearing (Chapter 8).

37 Clutch pedal stays on floor

1   Clutch release lever shaft binding in housing (Chapter 8).
2   Broken release bearing or bent lever (Chapter 8).

38 High pedal effort

1   Release lever shaft binding in housing (Chapter 8).
2   Pressure plate faulty (Chapter 8).
3   Incorrect size master or slave cylinder fitted (Chapter 8).

Manual transmission

39 Vibration

1   Rough wheel bearing (Chapters 1 and 10).
2   Damaged driveaxle (Chapter 8).
3   Out of round tires (Chapter 1).
4   Tire out of balance (Chapters 1 and 10).

40 Noisy in neutral with engine running

Damaged clutch release bearing (Chapter 8).

41 Noisy in one particular gear

1   Damaged or worn constant mesh gears (Chapter 7).
2   Damaged or worn synchronizers (Chapter 7).

42 Noisy in all gears

1   Insufficient lubricant (Chapter 7).
2   Damaged or worn bearings (Chapter 7).
3   Worn or damaged input gear shaft and/or output gear shaft (Chapter 7).

43 Slips out of gear

1   Worn or improperly adjusted linkage (Chapter 7).
2   Transmission loose on central tube (Chapter 7).
3   Shift linkage does not work freely, binds (Chapter 7).

44 Leaks lubricant

1   Excessive amount of lubricant in transmission (Chapters 1 and 7).
2   Loose or broken input shaft bearing retainer (Chapter 7).
3   Differential axleshaft seal seal damaged (Chapter 7).

Automatic transmission

Note: *Due to the complexity of the automatic transmission, it is difficult for the home mechanic to properly diagnose and service this component. For problems other than the following, the vehicle should be taken to a dealer or transmission shop.*

45 Fluid leakage

1   Automatic transmission fluid is a deep red color. Fluid leaks should not be confused with engine oil, which can easily be blown by air flow to the transmission.

2   To pinpoint a leak, first remove all built-up dirt and grime from the transmission housing with degreasing agents and/or steam cleaning. Then drive the vehicle at low speeds so air flow will not blow the leak far from its source. Raise the vehicle and determine where the leak is coming from. Common areas of leakage are:

   a) Pan (Chapters 1 and 7)
   b) Filler pipe (Chapter 7)
   c) Transmission oil lines (Chapter 7)
   d) Speedometer sensor (later models) (Chapter 7)
   e) Differential axleshaft flange seal (Chapter 7)

46 Transmission fluid brown or has a burned smell

Transmission fluid burned (Chapter 1).

47 General shift mechanism problems

1   Chapter 7 Part B deals with checking and adjusting the shift linkage on automatic transmissions. Common problems which may be attributed to poorly adjusted linkage are:

   a) Engine starting in gears other than Park or Neutral.
   b) Indicator on shifter pointing to a gear other than the one actually being used.
   c) Vehicle moves when in Park.

2   Refer to Chapter 7 Part B for the shift linkage adjustment procedure.

48 Transmission will not downshift with accelerator pedal pressed to the floor

Throttle pressure cable out of adjustment (Chapter 7).

49 Engine will start in gears other than Park or Neutral

Selector lever cable out of adjustment (Chapter 7).

50 Transmission slips, shifts roughly, is noisy or has no drive in forward or reverse gears

There are many probable causes for the above problems, but the home mechanic should be concerned with only one possibility — fluid level. Before taking the vehicle to a repair shop, check the level and condition of the fluid as described in Chapter 1. Correct the fluid level as necessary or change the fluid and filter if needed. If the problem persists, have a professional diagnose the probable cause.

Brakes

Note: *Before assuming that a brake problem exists, make sure that:*
   a) The tires are in good condition and properly inflated (Chapter 1).
   b) The front end alignment is correct (Chapter 10).
   c) The vehicle is not loaded with weight in an unequal manner.

51 Vehicle pulls to one side during braking

1   Incorrect tire pressures (Chapter 1).
2   Front end out of line (have the front end aligned).
3   Unmatched tires on same axle.

4   Restricted brake lines or hoses (Chapter 9).
5   Malfunctioning caliper assembly (Chapter 9).
6   Loose suspension parts (Chapter 10).
7   Loose calipers (Chapter 9).

52 Noise (high-pitched squeal when the brakes are applied)

Front and/or rear disc brake pads worn out. Replace pads with new ones immediately (Chapter 9).

53 Brake roughness or chatter (pedal pulsates)

1   Excessive lateral runout (Chapter 9).
2   Parallelism not within specifications (Chapter 9).
3   Defective rotor (Chapter 9).

54 Excessive brake pedal effort required to stop vehicle

1   Malfunctioning power brake booster (Chapter 9).
2   Partial system failure (Chapter 9).
3   Excessively worn pads (Chapter 9).
4   Piston in caliper stuck or sluggish (Chapter 9).
5   Brake pads contaminated with oil or grease (Chapter 9).
6   New pads installed and not yet seated. It will take a while for the new material to seat against the rotor.

55 Excessive brake pedal travel

1   Partial brake system failure (Chapter 9).
2   Insufficient fluid in master cylinder (Chapters 1 and 9).
3   Air trapped in system (Chapters 1 and 9).

56 Dragging brakes

1   Master cylinder pistons not returning correctly (Chapter 9).
2   Restricted brakes lines or hoses (Chapters 1 and 9).
3   Incorrect parking brake adjustment (Chapter 1).

57 Grabbing or uneven braking action

1   Malfunction of power brake booster unit (Chapter 9).
2   Binding brake pedal mechanism (Chapter 9).

58 Brake pedal feels spongy when depressed

1   Air in hydraulic lines (Chapter 9).
2   Master cylinder mounting bolts loose (Chapter 9).
3   Master cylinder defective (Chapter 9).

59 Brake pedal travels to the floor with little resistance

Little or no fluid in the master cylinder reservoir caused by leaking caliper piston(s), loose, damaged or disconnected brake lines (Chapter 9).

60 Parking brake does not hold

1   Parking brake linkage improperly adjusted (Chapters 1 and 9).
2   Parking brake lining worn (Chapter 9).

Suspension and steering systems

Note: *Before attempting to diagnose the suspension and steering systems, perform the following preliminary checks:*

1 Tires for wrong pressure and uneven wear.
2 Steering universal joint from the column to the rack and pinion for loose connectors or wear.
3 Front and rear suspension and the rack and pinion assembly for loose or damaged parts.
4 Out-of-round or out-of-balance tires, bent rims and loose and/or rough wheel bearings.

61 Vehicle pulls to one side

1 Mismatched or uneven tires (Chapter 10).
2 Broken or sagging springs (Chapter 10).
3 Front wheel or rear wheel alignment (Chapter 10).
4 Front brakes dragging (Chapter 9).

62 Abnormal or excessive tire wear

1 Front wheel or rear wheel alignment (Chapter 10).
2 Sagging or broken springs or torsion bars (Chapter 10).
3 Tire out of balance (Chapter 10).
4 Worn shock absorber (Chapter 10).
5 Overloaded vehicle.
6 Tires not rotated regularly.

63 Wheel makes a 'thumping' noise

1 Blister or bump on tire (Chapter 10).
2 Improper shock absorber action (Chapter 10).

64 Shimmy, shake or vibration

1 Tire or wheel out-of-balance or out-of-round (Chapter 10).
2 Loose, worn or out-of-adjustment wheel bearings (Chapters 1, 8 and 10).
3 Worn tie rod ends (Chapter 10).
4 Worn lower balljoints (Chapter 10).
5 Excessive wheel runout (Chapter 10).
6 Blister or bump on tire (Chapter 10).

65 Hard steering

1 Lack of lubrication at balljoints, tie rod ends and rack and pinion assembly (Chapter 10).
2 Front wheel alignment (Chapter 10).
3 Low tire pressure(s) (Chapters 1 and 10).

66 Poor returnability of steering to center

1 Lack of lubrication at balljoints and tie rod ends (Chapter 10).
2 Binding in balljoints (Chapter 10).
3 Binding in steering column (Chapter 10).
4 Lack of lubricant in rack and pinion assembly (Chapter 10).
5 Front wheel alignment (Chapter 10).

67 Abnormal noise at the front end

1 Lack of lubrication at balljoints and tie rod ends (Chapters 1 and 10).
2 Damaged shock absorber mounting (Chapter 10).
3 Worn control arm bushings or tie rod ends (Chapter 10).
4 Loose stabilizer bar (Chapter 10).

5 Loose wheel nuts (Chapters 1 and 10).
6 Loose suspension bolts (Chapter 10).

68 Wander or poor steering stability

1 Mismatched or uneven tires (Chapter 10).
2 Lack of lubrication at balljoints and tie rod ends (Chapters 1 and 10).
3 Worn shock absorbers (Chapter 10).
4 Loose stabilizer bar (Chapter 10).
5 Broken or sagging springs or torsion bars (Chapter 10).
6 Front or rear wheel alignment (Chapter 10).

69 Erratic steering when braking

1 Wheel bearings worn (Chapters 1, 8 and 10).
2 Broken or sagging springs or torsion bars (Chapter 10).
3 Leaking wheel cylinder or caliper (Chapter 10).
4 Warped rotors (Chapter 10).

70 Excessive pitching and/or rolling around corners or during braking

1 Loose stabilizer bar (Chapter 10).
2 Worn shock absorbers or mounting (Chapter 10).
3 Broken or sagging springs or torsion bars (Chapter 10).
4 Overloaded vehicle.

71 Suspension bottoms

1 Overloaded vehicle.
2 Worn shock absorbers (Chapter 10).
3 Incorrect, broken or sagging springs or torsion bars (Chapter 10).

72 Cupped tires

1 Front wheel or rear wheel alignment (Chapter 10).
2 Worn shock absorbers (Chapter 10).
3 Wheel bearings worn (Chapters 1, 8 and 10).
4 Excessive tire or wheel runout (Chapter 10).
5 Worn balljoints (Chapter 10).

73 Excessive tire wear on outside edge

1 Inflation pressures incorrect (Chapter 1).
2 Excessive speed in turns.
3 Front end alignment incorrect (excessive toe-in). Have professionally aligned.
4 Suspension arm bent or twisted (Chapter 10).

74 Excessive tire wear on inside edge

1 Inflation pressures incorrect (Chapter 1).
2 Front end alignment incorrect (toe-out). Have professionally aligned.
3 Loose or damaged steering components (Chapter 10).

75 Tire tread worn in one place

1 Tires out of balance.
2 Damaged or buckled wheel. Inspect and replace if necessary.
3 Defective tire (Chapter 1).

76   Excessive play or looseness in steering system

1   Wheel bearing(s) worn (Chapter 10).
2   Tie rod end loose or worn (Chapter 10).
3   Rack and pinion loose (Chapter 10).

77   Rattling or clicking noise in rack and pinion

1   Insufficient or improper lubricant in rack and pinion assembly (Chapter 10).
2   Rack and pinion attachment loose (Chapter 10).

Chapter 1   Tune-up and routine maintenance

Contents

Specifications

Recommended lubricants and fluids

Engine oil type	SF, SF/CC or SF/CD — consult a Porsche dealer for recommended brands
Engine oil viscosity	See chart below
Automatic transmission fluid	Dexron II ATF
Manual transmission oil	SAE 80W API Classification GL 4 gear lubricant
Differential (rear axle unit — automatic transmission) oil	SAE 90W gear oil
Engine coolant	Mixture of water and ethylene glycol phosphate-free antifreeze recommended for aluminum engines and radiators
Brake fluid	DOT 3 or 4 SAE Specification J1703 brake fluid
Clutch fluid	DOT 3 or 4 SAE Specification J1703 brake fluid
Power steering fluid	Dexron II ATF
Chassis lubrication	Multi-purpose lithium-base chassis grease

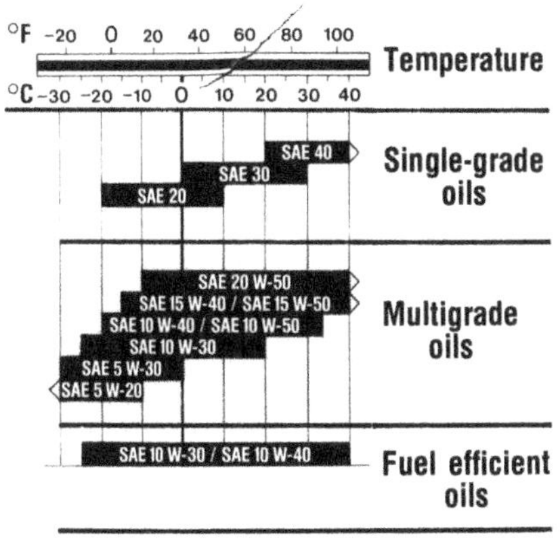

Capacities

Engine oil (approximate)	5.8 US qts (5.5 liters)
Coolant	2.2 US gal (8.5 liters)
Fuel tank	19.8 US gal (75 liters)
Automatic transmission	
fluid change only	3.0 US qt (2.8 liters)
complete system fill	6.9 US qt (6.5 liters)
Automatic transmission differential	1.05 US qt (1.0 liter)
Manual transmission	2.75 US qt (2.6 liters)

ENGINE OIL VISCOSITY

General

Ignition timing . . . .   Refer to *Vehicle Emission Control Information* label in the engine compartment

Spark plug type
    1983 through 1987 . . . .   Bosch WR8DS or Champion RN10GY
    2.7 liter WR7DC
    3.0 liter WR5DC
Spark plug gap . . . .   0.028 to 0.031 in (0.7 to 0.8 mm)
Cylinder numbers (front-to-rear)   1–2–3–4
Firing order   1-3-4-2
Engine idle speed . . . .   Not owner adjustable (See Section 28)
Alternator drivebelt deflection . . . .   0.200 in (5 mm)

Clutch

Clutch cylinder piston-to-edge of hole
    new clutch . . . .   0.71 in (18 mm)
    service limit . . . .   1.34 in (34 mm)

Brakes

Brake pad wear limit . . . .   0.080 (5/64) in (2 mm)

2479cc, 2681cc & 2969cc engines Firing order 1-3-4-2

Cylinder location and distributor rotation

Torque specifications

	Ft-lbs	Nm
Differential (rear axle – automatic transmission) fill plug	18 to 22	25 to 30
Engine block coolant drain plug	15	20
Spark plugs	18 to 22	25 to 30
Engine oil drain plug	44	60
Manual transmission check/fill plugs	18	25
Wheel lug nuts	96	130

1   Introduction

This chapter is designed to help the home mechanic maintain his or her vehicle for peak performance, economy, safety and long life.

On the following pages you will find a maintenance schedule along with sections which deal specifically with each item on the schedule. Included are visual checks, adjustments and item replacements.

Servicing your vehicle using the time/mileage maintenance schedule and the sequenced sections will give you a planned program of maintenance. Keep in mind that it is a full plan, and maintaining only a few items at the specified intervals will not give you the same results.

You will find as you service your vehicle that many of the procedures can, and should, be grouped together, due to the nature of the job at hand. Examples of this are as follows:

If the vehicle is raised for a chassis lubrication, for example, it is an ideal time for the following checks: exhaust system, suspension, steering and fuel system.

If the tires and wheels are removed, as during a routine tire rotation, check the brakes and wheel bearings at the same time.

If you must borrow or rent a torque wrench, service the spark plugs and check other components all in the same day to save time and money.

The first step of the maintenance plan is to prepare yourself before the actual work begins. Read through the appropriate Sections of this chapter for all work that is to be performed before you begin. Gather together all the necessary parts and tools. If it appears that you could have a problem during a particular job, don't hesitate to seek advice from your local parts man or dealer service department.

2   Maintenance schedule

The following recommendations are given with the assumption that the vehicle owner will be doing the maintenance or service work, as opposed to having a dealer service department do the work. The following are factory maintenance recommendations. However, the owner, interested in keeping his or her vehicle in peak condition at all times and with the vehicle's ultimate resale in mind, may want to perform many of these operations more often. Specifically, we would encourage the shortening of fluid and filter replacement intervals.

When the vehicle is new it may be wise to have the vehicle serviced initially by a factory authorized dealer service department to protect the factory warranty. In many cases the initial maintenance check is done at no cost to the owner. Check with your local dealer for additional information.

Every 250 miles or weekly, whichever comes first

Check the engine oil level (Section 4)
Check the engine coolant level (Section 4)
Check the windshield washer fluid level (Section 4)
Check the brake fluid and clutch fluid levels (Section 4)
Check the tires and tire pressures (Section 5)

Every 3000 miles or 3 months, whichever comes first

All items listed above plus:
Inspect and replace if necessary the windshield wiper blades (Section 6)
Check the cooling system (Section 7)
Check the power steering fluid level (Section 8)
Inspect and replace if necessary all underhood hoses (Section 9)
Check the automatic transmission fluid level (Section 10)
Change the engine oil and oil filter (Section 11)

Every 6000 miles or 6 months, whichever comes first

All items listed above plus:
Replace the air filter element (Section 12)
Check the engine drivebelts (Section 13)
Check and service the battery (Section 14)
Inspect the suspension and steering components (Section 15)
Inspect the exhaust system (Section 16)
Check the manual transmission oil level (Section 17)
Rotate the tires (Section 18)
Check the brakes (Section 19)
Inspect the driveaxle CV joints and boots for damage, wear and lubricant leakage (Section 20)
Inspect the fuel system (Section 21)
Check the front wheel bearings (Section 22)

27

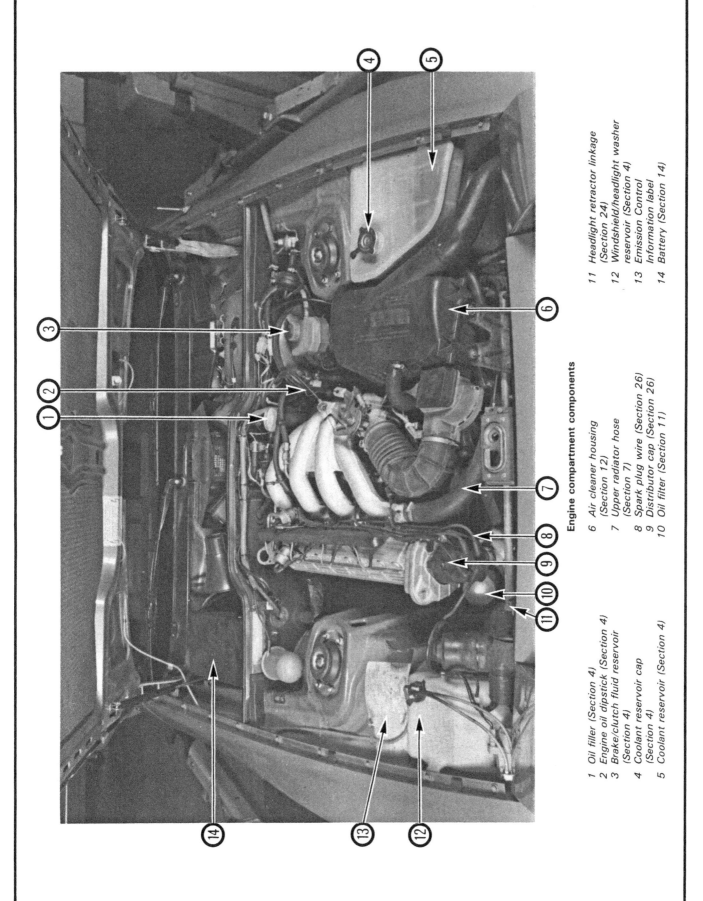

Engine compartment components

1 Oil filler (Section 4)
2 Engine oil dipstick (Section 4)
3 Brake/clutch fluid reservoir (Section 4)
4 Coolant reservoir cap (Section 4)
5 Coolant reservoir (Section 4)
6 Air cleaner housing (Section 12)
7 Upper radiator hose (Section 7)
8 Spark plug wire (Section 26)
9 Distributor cap (Section 26)
10 Oil filter (Section 11)
11 Headlight retractor linkage (Section 24)
12 Windshield/headlight washer reservoir (Section 4)
13 Emission Control Information label
14 Battery (Section 14)

Engine compartment underside components

1 Exhaust manifold and pipe
  (Section 16)
2 Lower radiator hose
  (Section 7)
3 Oil filter (Section 4)
4 Cooling fan (Chapter 3)

5 Drivebelt adjuster
  (Section 13)
6 Upper radiator hose
  (Section 7)
7 Air conditioning compressor
  drivebelt (Section 13)

8 Air conditioning compressor
  (Chapter 3)
9 Steering gear boot
  (Section 15)
10 Clutch inspection hole
  (Section 28)

29

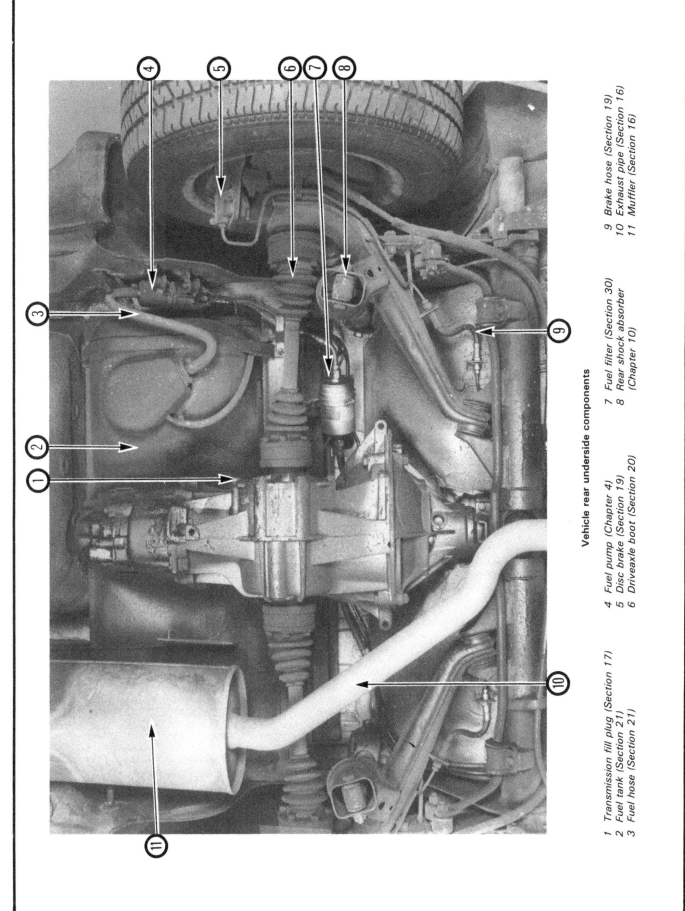

Vehicle rear underside components

1 Transmission fill plug (Section 17)
2 Fuel tank (Section 21)
3 Fuel hose (Section 21)
4 Fuel pump (Chapter 4)
5 Disc brake (Section 19)
6 Driveaxle boot (Section 20)
7 Fuel filter (Section 30)
8 Rear shock absorber (Chapter 10)
9 Brake hose (Section 19)
10 Exhaust pipe (Section 16)
11 Muffler (Section 16)

Every 7500 miles or 7.5 months

Have the camshaft drivebelt tension checked by a dealer (Section 31)
Have the balance shaft drivebelt tension checked by
   a dealer (Section 32)

Every 12,000 miles or 12 months, whichever comes first

All items listed above plus:
Inspect the evaporative emissions control system (Section 23)
Lubricate the headlight retractor linkage (Section 24)
Replace the spark plugs (Section 25)
Inspect the spark plug wires, distributor cap and rotor (Section 26)
Have the engine idle speed checked and adjusted if necessary
   by a dealer or a properly equipped shop (Section 27)
Check the clutch disc wear (Section 28)
Check the differential (rear axle) oil level (automatic
   transmission models) (Section 29)
Replace the fuel filter (Section 30)

Every 24,000 miles or 24 months, whichever comes first

All items listed above plus:
Service the cooling system (drain, flush and refill) (Section 33)
Change the automatic transmission fluid and clean the
   filter (Section 34)
Replace the brake fluid (Section 35)
Repack the front wheel bearings (Section 22)

Every 36,000 miles or 36 months, whichever comes first

All items listed above plus:
Change the manual transmission oil (Section 36)

3   Tune-up general information

The term *tune-up* is used in this manual to represent a combination
of individual operations rather than one specific procedure.

If, from the time the vehicle is new, the routine maintenance schedule
is followed closely and frequent checks are made of fluid levels and
high wear items, as suggested throughout this manual, the engine will
be kept in relatively good running condition and the need for additional
work will be minimized.

More likely than not, however, there will be times when the engine
is running poorly due to lack of regular maintenance. This is even more
likely if a used vehicle, which has not received regular and frequent
maintenance checks, is purchased. In such cases an engine tune-up
will be needed outside of the regular routine maintenance intervals.

The first step in any tune-up or engine diagnosis to help correct a
poor running engine would be a cylinder compression check. A check
of the engine compression will give valuable information regarding the
overall condition of many internal components and should be used as
a basis for tune-up and repair procedures. If, for instance, a compression
check indicates serious internal engine wear, a conventional tune-up
will not improve engine performance and would be a waste of time
and money. Due to its importance, compression checking should be
performed by someone who has the proper compression testing gauge
and who is knowledgeable with its use. Further information on compression testing can be found in Chapter 2 of this manual.

The following series of operations are those most often needed to
bring a generally poor running engine back into a proper state of tune.

Minor tune-up

Clean, inspect and check the battery
Check all engine-related fluids
Check and adjust the drivebelts
Replace the spark plugs

Inspect the distributor cap and rotor
Inspect the spark plug and coil wires
Have the ignition timing checked and adjusted
Check the air filter
Check all underhood hoses

Major tune-up

All operations listed under Minor Tune-up plus . . .
Check the cylinder compression
Have the camshaft and balance shaft drivebelts checked and,
   if necessary, adjusted
Check the EGR system
Check the ignition system
Check the charging system
Check the fuel system
Replace the air filter
Replace the distributor cap and rotor
Replace the spark plug wires
Check the cooling system

4   Fluid level checks

See illustrations 4.2, 4.4, 4.6, 4.9, 4.15, 4.18 and 4.21

Warning: *The electric cooling fan on these models can activate at any
time, even when the ignition is in the Off position. Disconnect the fan
motor or negative battery cable when working in the vicinity of the fan.*

Note: *The following are fluid level checks to be done on a 250 mile
or weekly basis. Additional fluid level checks can be found in specific
maintenance intervals which follow. Regardless of intervals, be alert
to fluid leaks under the vehicle which would indicate a fault to be corrected immediately.*

1   There are a number of components on a vehicle which rely on fluids
for lubrication and operation. During normal use of the vehicle, these
fluids are used up and must be replenished before damage occurs. See
Recommended lubricants and fluids at the front of this Chapter for the
specific fluid to be used when addition is required. When checking fluid
levels, be sure that the vehicle is on a level surface.

Engine oil

2   The engine oil level is checked with a dipstick **(see illustration)**.
The dipstick travels through a tube and into the oil pan at the bottom
of the engine.
3   The oil level should be checked before the vehicle has been driven,
or about 15 minutes after the engine has been shut off. If the oil is
checked immediately after driving the vehicle, some of the oil will remain in the upper engine components, producing an inaccurate reading
on the dipstick.
4   Pull the dipstick from the tube and wipe all the oil from the end
with a clean rag or paper towel. Insert the clean dipstick all the way
back into the tube, then pull it out again. Note the oil at the end of
the dipstick. At its highest point, the level should be between the upper and lower marks **(see illustration)**.
5   It takes approximately one quart of oil to raise the level from the
lower mark to the upper mark on the dipstick. Do not allow the level
to drop below the lower mark as engine damage due to oil starvation
may occur. On the other hand, do not overfill the engine by adding
oil above the upper mark since it may result in oil fouled spark plugs,
oil leaks or oil seal failures.
6   Oil is added to the engine after removing a twist off cap marked
Oil located at the rear of the engine, adjacent to the firewall **(see illustration)**. An oil can spout or funnel will reduce spills.
7   Checking the oil level can be an important preventive maintenance
step. If you find the oil level dropping abnormally, it is an indication
of oil leakage or internal engine wear which should be corrected. If
there are water droplets in the oil, or if it is milky looking, component
failure is indicated and the engine should be checked immediately. The
condition of the oil can also be checked along with the level. With the
dipstick removed from the engine, take your thumb and index finger
and wipe the oil up the dipstick, looking for small dirt or metal particles
which will cling to the dipstick. This is an indication that the oil should
be drained and fresh oil added (Section 11).

Engine coolant

8   All vehicles covered by this manual are equipped with a pressurized coolant recovery system. A white coolant reservoir attached to the left inner fender panel is connected by hoses to the radiator. As the engine heats up during operation, coolant is forced from the radiator, through the connecting tubes and into the reservoir. As the engine cools, the coolant is automatically drawn back into the radiator to keep the level correct.

9   The coolant level should be checked when the engine is hot. Note the level of fluid in the reservoir, which should be at or near the Maximum mark on the side of the reservoir **(see illustration)**. **Warning:** *Under no circumstances should the coolant recovery reservoir cap be removed when the system is hot, because escaping steam and scalding coolant could cause serious personal injury.*

10   After it is apparent that no further boiling is occurring in the reservoir, wrap a thick cloth around the cap and turn it to the first stop. If any steam escapes, wait until the system has cooled further, then remove the cap.

11   If only a small amount of coolant is required to bring the system up to the proper level, plain water can be used. However, to maintain the proper antifreeze/water mixture in the system, both should be mixed together to replenish a low level. High quality antifreeze offering protection to − 20°F should be mixed with water in the proportion specified on the container. Do not allow antifreeze to come in contact with your skin or painted surfaces of the vehicle. Flush contacted areas immediately with plenty of water.

12   Coolant should be added to the reservoir until it reaches the Maximum mark.

13   As the coolant level is checked, note the condition of the coolant.

It should be relatively clear. If it is brown or a rust color, the system should be drained, flushed and refilled (Section 33).

14   If the cooling system requires repeated additions to maintain the proper level, have the coolant reservoir cap checked for proper sealing ability. Also check for leaks in the system from cracked hoses, loose hose connections, leaking gaskets, etc.

Windshield/headlight washer fluid

15   Fluid for the windshield and optional headlight washer system is located in a plastic reservoir behind the right front headlight **(see illustration)**. The reservoir should be kept no more than 2/3-full to allow for expansion should the fluid freeze. The use of a wiper fluid additive, available at auto parts stores, will help lower the freezing point of the fluid and will result in better cleaning of the windshield surface. **Caution:** *Do not use cooling system antifreeze — it will cause damage to the vehicle's paint.*

16   To help prevent icing in cold weather, warm the windshield with the defroster before using the washer.

Battery electrolyte

17   Some of these vehicles are equipped with a maintenance-free battery which is permanently sealed (except for vent holes). It has no filler caps and does not require adding water at any time.

18   If equipped with a conventional battery, the caps on the top of the battery should be removed periodically to check for a low electrolyte level **(see illustration)**. This check is more critical during the warm summer months.

19   Remove each of the caps and add distilled water to bring the level of each cell to the split ring in the opening.

4.2   The engine oil level dipstick is located at the left rear corner of the engine

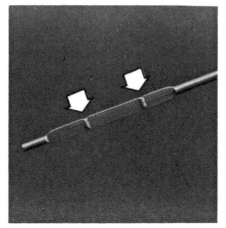
4.4   The oil level must be maintained between the marks on the dipstick (arrows)

4.6   Oil is added to the engine through a filler cap located at the rear of the engine

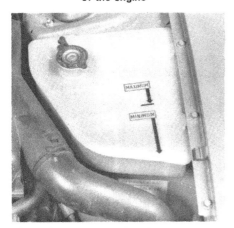
4.9   The coolant level (which should be kept between the Maximum and Minimum marks) is checked at the reservoir, as these models don't have a cap on the radiator

4.15   The windshield washer fluid reservoir is located at the right front corner of the engine compartment

4.18   The battery electrolyte level can be checked after unscrewing the caps (conventional batteries only)

Brake and clutch fluid

20   The brake and clutch master cylinder is mounted on the front of the power booster unit in the engine compartment.

21   The translucent plastic reservoir allows the fluid level to be checked without removing the cap **(see illustration)**. If a low level is indicated, be sure to wipe the top of the reservoir with a clean rag to prevent contamination of the brake and/or clutch system before removing the cap.

22   When adding fluid, pour it carefully into the reservoir, taking care not to spill any onto surrounding painted surfaces. Be sure the specified fluid is used, since mixing different types of brake fluid can cause damage to the system. See *Recommended lubricants and fluids* near the front of this Chapter or your owner's manual.

23   At this time the fluid and master cylinder can be inspected for contamination. The system should be drained and refilled if deposits, dirt particles or water droplets are seen in the fluid.

24   After filling the reservoir to the correct level, make sure the cap is properly installed to prevent fluid leakage.

25   The brake fluid in the master cylinder will drop slightly as the brake shoes or pads at each wheel wear down during normal operation. If the master cylinder requires repeated replenishing to keep it at the proper level, this is an indication of leakage in the brake system, which should be corrected immediately. Check all brake lines and connections, along with the wheel cylinders and booster (see Section 19 for more information).

26   If, upon checking the master cylinder fluid level, you discover that the reservoir is empty or nearly empty, the brake system should be bled (Chapter 9).

4.21   The translucent plastic brake and clutch fluid reservoir allows the fluid level to be checked without removing the cap

---

5   Tire and tire pressure checks

Refer to illustration 5.3

1   Periodically inspecting the tires may not only prevent you from being stranded with a flat tire, but can also give you clues as to possible problems with the steering and suspension systems before major damage occurs.

2   Proper tire inflation adds miles to the lifespan of the tires, allows the vehicle to achieve maximum miles per gallon figures and contributes to the overall quality of the ride.

3   When inspecting the tires, first check for wear on the tread. Irregularities in the tread pattern (cupping, flat spots, more wear on one side than the other) are indications of front end alignment and/or balance problems **(see illustration)**. If any of these conditions are noted, take the vehicle to a repair shop to correct the problem.

4   Check the tread area for cuts and punctures. Many times a nail or tack will embed itself in the tire tread and yet the tire may hold air pressure for a period of time. In most cases, a repair shop or gas station can repair the punctured tire.

5   It is important to check the sidewalls of the tires, both inside and outside. Check for deteriorated rubber, cuts, and punctures. Inspect the inboard side of the tire for signs of brake fluid leakage, indicating

Condition	Probable cause	Corrective action	Condition	Probable cause	Corrective action
Shoulder wear	• Underinflation (both sides wear) • Incorrect wheel camber (one side wear) • Hard cornering • Lack of rotation	• Measure and adjust pressure. • Repair or replace axle and suspension parts. • Reduce speed. • Rotate tires.	Feathered edge **Toe wear**	• Incorrect toe	• Adjust toe-in.
Center wear	• Overinflation • Lack of rotation	• Measure and adjust pressure. • Rotate tires.	**Uneven wear**	• Incorrect camber or caster • Malfunctioning suspension • Unbalanced wheel • Out-of-round brake drum • Lack of rotation	• Repair or replace axle and suspension parts. • Repair or replace suspension parts. • Balance or replace. • Turn or replace. • Rotate tires.

5.3   This chart will help you determine the condition of your tires, the probable cause(s) of abnormal wear and the corrective action necessary

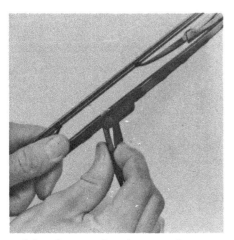

6.6a   Squeeze the wiper arm clip . . .

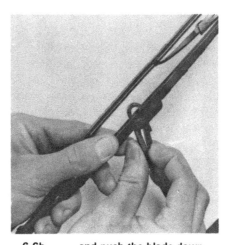

6.6b   . . . and push the blade down over the arm

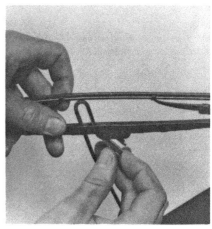

6.7   Hold the clip out of the way and lift the blade over the arm hook

that a thorough brake inspection is needed immediately.

6   Incorrect tire pressure cannot be determined merely by looking at the tire. This is especially true for radial tires. A tire pressure gauge must be used. If you do not already have a reliable gauge, it is a good idea to purchase one and keep it in the glovebox. Built-in pressure gauges at gas stations are often inaccurate.

7   Always check tire inflation when the tires are cold. Cold, in this case, means the vehicle has not been driven more than one mile after sitting for three hours or more. It is normal for the pressure to increase four to eight pounds when the tires are hot.

8   Unscrew the valve cap protruding from the wheel or hubcap and press the gauge firmly onto the valve. Note the reading on the gauge and compare the figure to the recommended tire pressure listed on the inside of the fuel filler cap door.

9   Check all tires and add air as necessary to bring them up to the recommended pressure levels. Be sure to reinstall the valve caps, which will keep dirt and moisture out of the valve stem mechanism.

6   Wiper blade inspection and replacement

Refer to illustrations 6.6a, 6.6b and 6.7

1   The windshield wiper and blade assembly should be inspected periodically for damage, loose components and cracked or worn blade elements.

2   Road film can build up on the wiper blades and affect their efficiency, so they should be washed regularly with a mild detergent solution.

3   The action of the wiping mechanism can loosen the bolts, nuts and fasteners, so they should be checked and tightened, as necessary, at the same time the wiper blades are checked.

4   If the wiper blade elements are cracked, worn or warped, they should be replaced with new ones.

5   Cycle the wiper up from the parked position and shut off the key.

6   Lift the arm assembly away from the glass for clearance, squeeze the wiper arm clip and push the wiper blade assembly down over the arm hook to disengage it **(see illustrations)**.

7   Hold the clip out of the way and lift the blade assembly over the hook and off the arm **(see illustration)**.

8   Installation is the reverse of removal.

7   Cooling system check

Refer to illustration 7.4

1   Many major engine failures can be attributed to a faulty cooling system. If the vehicle is equipped with an automatic transmission, the cooling system also cools the transmission fluid and thus plays an important role in prolonging transmission life.

2   The cooling system should be checked with the engine cold. Do

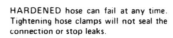

ALWAYS CHECK hose for chafed or burned areas that may cause an untimely and costly failure.

SOFT hose indicates inside deterioration. This deterioration can contaminate the cooling system and cause particles to clog the radiator.

HARDENED hose can fail at any time. Tightening hose clamps will not seal the connection or stop leaks.

SWOLLEN hose or oil soaked ends indicate danger and possible failure from oil or grease contamination. Squeeze the hose to locate cracks and breaks that cause leaks.

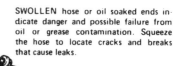

7.4   Hoses, like drivebelts, have a habit of failing at the worst possible time — to prevent the inconvenience of a blown radiator or heater hose, inspect them carefully as shown here

this before the vehicle is driven for the day or after it has been shut off for at least three hours.

3   Remove the reservoir cap and thoroughly clean the cap, inside and out, with clean water. Also clean the filler neck on the reservoir. All traces of corrosion should be removed. The coolant inside the reservoir should be relatively transparent. If it is rust-colored, the system should be drained and refilled (Section 33).

4   Carefully check the large upper and lower radiator hoses along with the smaller diameter heater hoses which run from the engine to the firewall. Inspect each hose along its entire length, replacing any hose which is cracked, swollen or shows signs of deterioration **(see illustration)**. Cracks may become more apparent if the hose is squeezed.

5   Make sure that all hose connections are tight. A leak in the cooling system will usually show up as white or rust colored deposits on the areas adjoining the leak.

6   Use compressed air or a soft brush to remove bugs, leaves, etc. from the front of the radiator or air conditioning condenser. Be careful not to damage the delicate cooling fins or cut yourself on them.

7   Every other inspection, or at the first indications of a fault in the cooling system, have the cap and system pressure tested. If you do not have a pressure tester, most gas stations and repair shops will do this for a minimal charge.

8   Power steering fluid level check

Warning: *The electric cooling switch on these models can activate at any time, even when the ignition switch is in the Off position. Disconnect the fan motor or negative battery cable when working in the vicinity of the fan.*

1   Unlike manual steering, the power steering system relies on fluid which may, over a period of time, require replenishing.

2   The fluid reservoir for the power steering pump is located behind the radiator near the front of the engine.

3   For the check, the front wheels should be pointed straight ahead and the engine should be off.

4   Use a clean rag to wipe off the reservoir cap and the area around the cap. This will help prevent any foreign matter from entering the reservoir during the check.

5   Run the engine until it is at normal operating temperature.

6   Remove the dipstick, wipe it off with a clean rag, reinsert it, then withdraw it and read the fluid level. The level should be between the upper and lower marks on the dipstick.

7   If additional fluid is required, pour the specified type directly into the reservoir, using a funnel to prevent spills.

8   If the reservoir requires frequent addition of fluid, all power steering hoses, hose connections, the power steering pump and the rack and pinion assembly should be carefully checked for leaks.

9   Underhood hose check and replacement

Warning: *The electric cooling fan on these models can activate at any time, even when the ignition is in the Off position. Disconnect the fan motor or negative battery cable when working in the vicinity of the fan. Replacement of air conditioning hoses must be left to a dealer or air conditioning specialist who has the proper equipment to depressurize the system safely. Never remove air conditioning components or hoses until the system has been depressurized.*

General

1   High temperatures under the hood can cause the deterioration of the rubber and plastic hoses used for engine, accessory and emission systems operation. Periodic inspection should be made for cracks, loose clamps, material hardening and leaks.

2   Information specific to the cooling system hoses can be found in Section 7.

3   Some, but not all, hoses are attached to the fittings with clamps. Where clamps are used, check to be sure they are tight. Where clamps are not used, make sure the hose has not expanded and/or hardened where it slips over the fitting, allowing it to leak.

Vacuum hoses

4   It is quite common for vacuum hoses, especially those in the emissions system, to be color coded or identified by colored stripes molded into the hose. Various systems require hoses with different wall thicknesses, collapse resistance and temperature resistance. When replacing hoses be sure to use the same hose material for the new one.

5   Often the only effective way to check a hose is to remove it completely from the vehicle. Where more than one hose is removed, be sure to label the hoses and fittings to ensure proper reattachment.

6   When checking vacuum hoses, be sure to include any plastic T-fittings in the check. Check the fittings for cracks and the hose where it fits over the fitting for enlargement, which could cause leakage.

7   A small piece of vacuum hose (1/4-inch inside diameter) can be used as a stethoscope to detect vacuum leaks. Hold one end of the hose to your ear and probe around vacuum hoses and fittings, listening for the hissing sound characteristic of a vacuum leak. **Warning:** *When probing with the vacuum hose stethoscope, be careful not to contact moving engine components such as drivebelts, the cooling fan, etc.*

Fuel hose

Warning: *There are certain precautions which must be taken when inspecting or servicing fuel system components. Work in a well ventilated area and do not allow open flames (cigarettes, appliance pilot lights, etc.) or bare light bulbs near the work area. Mop up any spills immediately and do not store fuel soaked rags where they could ignite.*

8   The fuel lines are under some pressure even with the engine off. The system must be depressurized (Chapter 4) before opening any fuel lines.

9   Check all rubber fuel hoses for deterioration and abrasions. Check especially for cracks in areas where the hose bends and just before clamping points, such as where a hose attaches to the fuel pump, fuel filter and fuel injection components.

10   High quality fuel hose, usually identified by the word *Fluroelastomer* printed on the hose, should be used for replacement. Under no circumstances should unreinforced vacuum line, clear plastic tubing or water hose be used for fuel hose replacement.

Metal lines

11   Metal lines are used in the fuel and brake systems. Check carefully to be sure each line has not been bent and crimped and that cracks have not started in the line, particularly in the area of bends.

12   If a section of metal fuel line must be replaced, only seamless steel tubing should be used, since copper and aluminum tubing do not have the strength necessary to withstand normal engine operating vibration.

13   Check the metal brake lines where they enter the master cylinder and brake proportioning unit (if used) for cracks in the lines and loose fittings. Any sign of brake fluid leakage calls for an immediate thorough inspection of the brake system.

10   Automatic transmission fluid level check

1   The fluid level in the automatic transmission should be carefully maintained. A low fluid level can lead to slipping or loss of drive, while overfilling can cause foaming and loss of fluid.

2   After the vehicle has been driven to warm up the transmission, raise the vehcle and support it securely on jackstands.

3   With the engine idling, check the fluid level in the reservoir, located on the side of the transmission, to make sure it is between the Max and Min marks.

4   If it is necessary to add fluid, unscrew the reservoir cap and add just enough of the recommended fluid to raise the level to just under the Max mark.

5   Screw the cap on securely and lower the vehicle.

11   Oil and oil filter change

Refer to illustrations 11.3 and 11.14

1   Frequent oil changes may be the best form of preventive maintenance available to the home mechanic. When engine oil ages, it becomes diluted and contaminated, which leads to premature engine wear.

2   Although some sources recommend oil filter changes every other oil change, we feel that the minimal cost of an oil filter and the relative ease with which it is installed dictate that a new filter be used whenever the oil is changed.

3   Gather together all necessary tools and materials before beginning the procedure **(see illustration)**.

4   In addition, you should have plenty of clean rags and newspapers handy to mop up any spills. Access to the underside of the vehicle is greatly improved if the vehicle can be lifted on a hoist, driven onto ramps or supported by jackstands. **Warning:** *Do not work under a vehicle which is supported only by a bumper, hydraulic or scissors-type jack.*

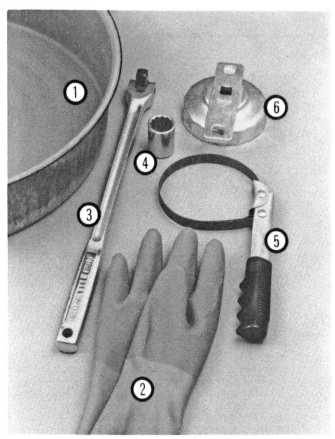

11.3   These tools are required when changing the
engine oil and filter

1   **Drain pan** — *It should be fairly shallow in depth, but
wide in order to prevent spills*
2   **Rubber gloves** — *When removing the drain plug and
filter it is inevitable that you will get oil on your hands
(the gloves will prevent burns)*
3   **Breaker bar** — *Sometimes the oil drain plug is pretty
tight and a long breaker bar is needed to loosen it*
4   **Socket** — *To be used with the breaker bar or a ratchet
(must be the correct size to fit the drain plug)*
5   **Filter wrench** — *This is a metal band-type wrench,
which requires clearance around the filter to be effective*
6   **Filter wrench** — *This type fits on the bottom of the
filter and can be turned with a ratchet or breaker bar
(different size wrenches are available for different types
of filters)*

5   If this is your first oil change, get under the vehicle and locate the
oil drain plug. The engine and exhaust components will be warm during
the actual work, so figure out any potential problems before starting
the engine.
6   Warm the engine to normal operating temperature. If the new oil
or any tools are needed, use this warm-up time to gather everything
necessary for the job. The correct type of oil for your application can
be found in *Recommended lubricants and fluids* at the beginning of
this chapter.
7   With the engine oil warm (warm engine oil will drain better and
more built-up sludge will be removed with the oil), raise and support
the vehicle. Make sure it is safely supported.
8   Move all necessary tools, rags and newspapers under the vehicle.
Position the drain pan under the drain plug. Keep in mind that the oil
will initially flow from the engine with some force, so locate the drain
pan accordingly.
9   Being careful not to touch any of the hot exhaust components,
use the wrench to remove the drain plug at the bottom of the oil pan.

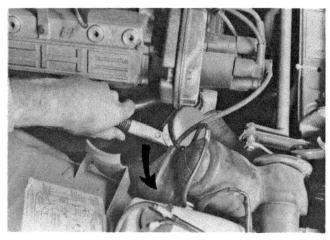

11.14   A strap-type wrench should be used to loosen the
oil filter, which can be reached from under the hood

Depending on how hot the oil has become, you may want to wear
gloves while unscrewing the plug the final few turns.
10   Allow the old oil to drain into the pan. It may be necessary to move
the pan farther under the engine as the oil flow slows to a trickle.
11   After all the oil has drained, wipe off the drain plug with a clean
rag. Small metal particles may cling to the plug and would immediately
contaminate the new oil.
12   Clean the area around the drain plug opening and reinstall the plug.
Tighten the plug securely with the wrench. If a torque wrench is available, use it to tighten the plug.
13   Move the drain pan into position under the oil filter.
14   Use the filter wrench to loosen the oil filter. Chain or metal band
filter wrenches may distort the filter canister, but this is of no concern
as the filter will be discarded anyway **(see illustration)**.
15   Sometimes the oil filter is on so tight it cannot be loosened. As
a last resort, you can punch a metal bar or long screwdriver directly
through the side of the canister and use it as a T-bar to turn the filter.
If so, be prepared for oil to spurt out of the canister as it is punctured.
16   Completely unscrew the old filter. Empty the oil inside the filter
into the drain pan.
17   Compare the old filter with the new one to make sure they are the
same type.
18   Use a clean rag to remove all oil, dirt and sludge from the area where
the oil filter mounts to the engine. Check the old filter to make sure
the rubber gasket is not stuck to the engine mounting surface. If the
gasket is stuck to the engine (use a flashlight if necessary), remove it.
19   Apply a light coat of oil to the rubber gasket on the new oil filter.
20   Attach the new filter to the engine, following the tightening directions printed on the filter canister or packing box. Most filter manufacturers recommend against using a filter wrench due to the possibility
of overtightening and damage to the seal.
21   Remove all tools, rags, etc. from under the vehicle, being careful
not to spill the oil in the drain pan, then lower the vehicle.
22   Move to the engine compartment and locate the oil filler cap.
23   If an oil can spout is used, push the spout into the top of the oil
can and pour the fresh oil through the filler opening. A funnel may also
be used.
24   Pour the specified amount of oil into the engine. Wait a few minutes
to allow the oil to drain into the pan, then check the level on the oil
dipstick (see Section 4 if necessary). If the oil level is at or near the upper
mark on the dipstick, start the engine and allow the new oil to circulate.
25   Run the engine for only about a minute and then shut it off. Immediately look under the vehicle and check for leaks at the oil pan drain
plug and around the oil filter. If either is leaking, tighten with a bit more
force.
26   With the new oil circulated and the filter now completely full,
recheck the level on the dipstick and add enough oil to bring the level
to the upper mark on the dipstick.
27   During the first few trips after an oil change, make it a point to
check frequently for leaks and proper oil level.

12.3   Lift up on the air filter cover while pulling the air intake out of the fender mount

12.4   Be sure to note which side of the filter element faces down before removing the element from the housing

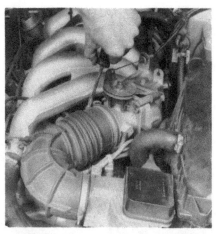

12.8   Use a screwdriver to tighten the air cleaner hose/duct clamps

SMALL CRACKS

GREASE

GLAZED

ALWAYS CHECK the underside of the belt.

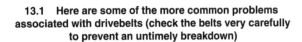

13.1   Here are some of the more common problems associated with drivebelts (check the belts very carefully to prevent an untimely breakdown)

13.4a   Push firmly on the belt to check the deflection

28   The old oil drained from the engine cannot be reused in its present state and should be disposed of. Oil reclamation centers, auto repair shops and gas stations will normally accept the oil, which can be refined and used again. After the oil has cooled, it can be drained into a suitable container (capped plastic jugs, topped bottles, milk cartons, etc.) for transport to one of these disposal sites.

12   Air filter replacement

Refer to illustrations 12.3, 12.4 and 12.8

1   At the specified intervals, the air filter should be replaced with a new one. A thorough program of preventive maintenance would call for the filter to be inspected between changes.

2   The air filter is located inside the air cleaner housing at the left front corner of the engine on models through 1988. On 1989 models, the housing is located beneath the front end panel, between the headlights. Remove the bolts, nuts and screws and slide the front end panel forward to remove it. **Note:** *You must raise the headlights to access some of these fasteners.*

3   Loosen the breather hose clamp and pull off the hose. Remove the screws around the top of the air cleaner assembly, lift the cover off the housing and disengage the rubber air intake from the fender mount **(see illustration)**.

4   Lift the air filter element out of the housing, noting the direction in which it is installed **(see illustration)**.

5   Wipe out the inside of the air cleaner housing with a clean rag.

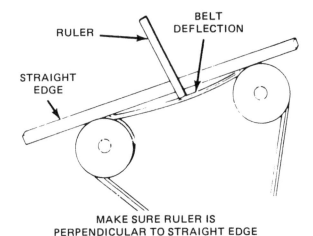

RULER

BELT DEFLECTION

STRAIGHT EDGE

MAKE SURE RULER IS PERPENDICULAR TO STRAIGHT EDGE

13.4b   Measuring drivebelt deflection with a straightedge and ruler

13.6   After the locknuts are loosened and backed off, turn the pressure rod to adjust the drivebelt tension

6   Place the new filter into the air cleaner housing. Make sure it seats properly in the bottom of the housing.
7   Install the cover on the housing, making sure the rubber air intake hose fits snugly in the fender mount, tighten the screws securely and connect the breather hose.
8   Check all of the air cleaner housing and hose fasteners to make sure there are no air leaks **(see illustration)**.

---

13   Drivebelt check, adjustment and replacement

Refer to illustrations 13.1, 13.4a, 13.4b and 13.6

Check

1   The drivebelts are located at the front of the engine and play an important role in the overall operation of the engine and related components. Due to their function and material make up, the belts are prone to failure after a period of time and should be inspected and adjusted periodically to prevent major engine damage **(see illustration)**.
2   The number of belts used on a particular vehicle depends on the accessories installed. Drivebelts are used to turn the alternator, power steering pump and air conditioning compressor. Depending on the pulley arrangement, a single belt may be used to drive more than one of these components.
3   With the engine off, open the hood and locate the various belts at the front of the engine. Using your fingers (and a flashlight, if necessary), move along the belts checking for cracks and separation of the belt plies. Also check for fraying and glazing, which gives the belt a shiny appearance. Both sides of the belt should be inspected, which means you will have to twist the belt to check the underside.
4   The tension of each belt is checked by pushing on the belt at a distance halfway between the pulleys. Push firmly with your thumb and see how much the belt moves (deflects) **(see illustrations)**. Measure the deflection with a ruler and compare the amount of deflection to the Specifications at the front of this Chapter.

Adjustment

5   If it is necessary to adjust the belt tension, either to make the belt tighter or looser, it is done by moving the belt-driven accessory on the bracket. On the power steering pump it will be necessary to loosen the pump pivot bolts prior to adjustment and retighten them after adjustment.
6   Each component is equipped with a pressure rod for drivebelt adjustment. Loosen the locknuts at each end of the pressure rod and turn the rod until the proper drivebelt tension is obtained **(see illustration)**.
7   Check the belt tension. If it is correct, tighten the locknuts.

Replacement

8   To replace a belt, follow the instructions above for adjustment, but completely remove the belt from the pulleys.
9   In some cases you will have to remove more than one belt because of their arrangement on the front of the engine. Due to this and the

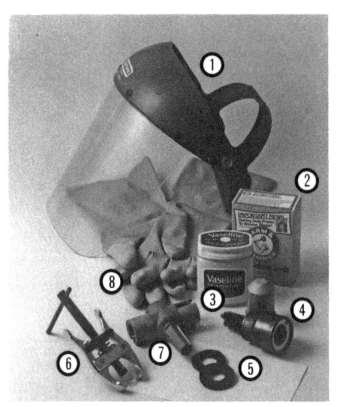

14.1   Tools and materials required for battery maintenance

1   **Face shield/safety goggles** — *When removing corrosion with a brush, the acidic particles can easily fly up into your eyes*
2   **Baking soda** — *A solution of baking soda and water can be used to neutralize corrosion*
3   **Petroleum jelly** — *A layer of this on the battery posts will help prevent corrosion*
4   **Battery post/cable cleaner** — *This wire brush cleaning tool will remove all traces of corrosion from the battery posts and cable clamps*
5   **Treated felt washers** — *Placing one of these on each post, directly under the cable clamps, will help prevent corrosion*
6   **Puller** — *Sometimes the cable clamps are very difficult to pull off the posts, even after the nut/bolt has been completely loosened. This tool pulls the clamp straight up and off the post without damage*
7   **Battery post/cable cleaner** — *Here is another cleaning tool which is a slightly different version of number 4 above, but it does the same thing*
8   **Rubber gloves** — *Another safety item to consider when servicing the battery; remember that's acid inside the battery!*

fact that belts will tend to fail at the same time, it is a good idea to replace all belts at the same time. Mark each belt and its appropriate pulley groove so all replacement belts can be installed in their proper positions.
10   When replacing belts, it is a good idea to take the old belts with you to the parts store in order to make a direct comparison for length, width and design.

---

14   Battery check and maintenance

Refer to illustrations 14.1, 14.5a, 14.5b, 14.5c and 14.5d

1   Battery maintenance is an important procedure which will help ensure that you are not stranded because of a dead battery. Several tools are required for this procedure **(see illustration)**.
2   **Warning:** *Hydrogen gas in small quantities is present in the area*

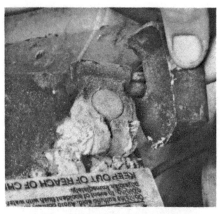

14.5a Battery terminal corrosion usually appears as light, fluffy powder

14.5b Removing the cable from a battery post with a wrench — sometimes a special battery pliers is required for this procedure if corrosion has caused deterioration of the nut hex (always remove the ground cable first and hook it up last!)

14.5c Regardless of the type of tool used to clean the battery posts, a clean, shiny surface should be the result

of the cap vents on batteries, so keep lighted tobacco and open flames or sparks away from them.

3 The external condition of the battery should be inspected periodically for damage such as a cracked case or cover.

4 Check the tightness of the battery cable clamps to ensure good electrical connections and check the entire length of each cable for cracks and frayed conductors.

5 If corrosion (visible as white, fluffy deposits) is evident, remove the cables from the terminals, clean them with a battery brush and reinstall the cables **(see illustrations)**. Corrosion can be kept to a minimum by installing specially treated washers available at auto parts stores or by applying a layer of petroleum jelly or grease to the terminals and cable clamps after they are assembled.

6 Make sure that the rubber protector (if so equipped) over the positive terminal is not torn or missing. It should completely cover the terminal.

7 Make sure that the battery carrier is in good condition and that the hold-down clamp bolts are tight. If the battery is removed from the carrier, make sure that no parts remain in the bottom of the carrier when the battery is reinstalled. When reinstalling the hold-down clamp bolts, do not overtighten them.

8 Corrosion on the hold-down components, battery case and surrounding areas may be removed with a solution of water and baking soda, but take care to prevent any solution from coming in contact with your eyes, skin or clothes. Protective gloves should be worn. Thoroughly wash all cleaned areas with plain water.

9 Any metal parts of the vehicle damaged by corrosion should be covered with a zinc-based primer then painted.

10 Further information on the battery, charging and jumpstarting can be found in Chapter 5 and at the front of this manual.

14.5d When cleaning the cable clamps, all corrosion must be removed (the inside of the clamp is tapered to match the taper on the post, so don't remove too much material)

be found in Chapter 10.

4 Raise the front end of the vehicle and support it securely on jackstands placed under the frame rails. Because of the work to be done, make sure the vehicle is safely supported.

5 Check the front wheel bearings (see Section 22).

6 From under the vehicle check for loose bolts, broken or disconnected parts and deteriorated rubber bushings on the suspension components, balljoints and tie-rod joints **(see illustrations)**.

7 Look for grease or fluid leaking from the steering assembly **(see illustration)**. Check the power steering hoses and connections for leaks.

8 Have an assistant turn the steering wheel from side-to-side and check the steering components for free movement, chafing and binding. If the steering does not react to the movement of the steering wheel, try to determine where the slack is located.

9 Lower the vehicle. With the vehicle weight resting on the suspension, pry up and down on the balljoints to check for looseness, indicating excessive wear **(see illustrations)**.

---

15 Suspension and steering check

Refer to illustrations 15.6a, 15.6b, 15.7, 15.9a and 15.9b

1 Raise the front of the vehicle periodically and visually check the suspension and steering components for wear.

2 Indications of a fault in these systems are excessive play in the steering wheel before the front wheels react, excessive sway around corners, body movement over rough roads or binding at some point as the steering wheel is turned.

3 Before the vehicle is raised for inspection, test the shock absorbers by pushing down to rock the vehicle at each corner. If you push down and the vehicle does not come back to a level position within one or two bounces, the shocks/struts are worn and must be replaced. As this is done, check for squeaks and noises coming from the suspension components. Additional information on suspension components can

---

16 Exhaust system check

Refer to illustrations 16.2a and 16.2b

1 With the engine cold (at least three hours after the vehicle has been driven), check the complete exhaust system from its starting point at

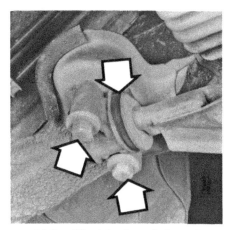

15.6a   Check for loose bolts and cracked bushings (arrows) and . . .

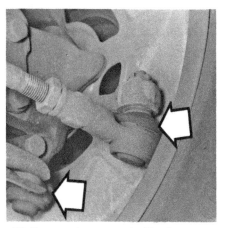

15.6b   . . . for torn or leaking boots on the tie-rods and balljoints (arrows)

15.7   Pull the steering boots back and check for oil, indicating a leaking seal in the rack and pinion

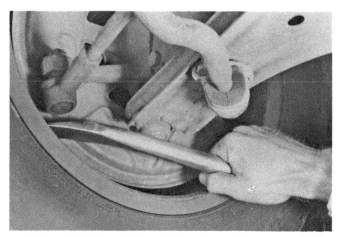

15.9a   With the vehicle weight resting on the suspension, pry up on the balljoint with a bar . . .

15.9b   . . . and then down to check for looseness, indicating wear

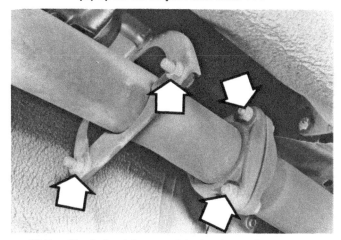

16.2a   Check the tightness of the bracket and pipe bolts and nuts . . .

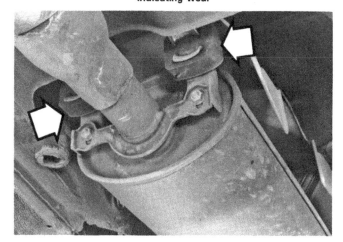

16.2b   . . . and look for cracked rubber hangers

the engine to the end of the tailpipe. This should be done on a hoist where unrestricted access is available.

2   Check the pipes and connections for signs of leakage and/or corrosion indicating a potential failure **(see illustration)**. Make sure that all brackets and hangers are in good condition and tight **(see illustration)**.

3   At the same time, inspect the underside of the body for holes, corrosion, open seams, etc. which may allow exhaust gases to enter the passenger compartment. Seal all body openings with silicone or body putty.

4   Rattles and other noises can often be traced to the exhaust system, especially the mounts and hangers. Try to move the pipes, muffler and catalytic converter. If the components can come in contact with the body or suspension parts, secure the exhaust system with new mounts.

5   Check the running condition of the engine by inspecting inside the end of the tailpipe. The exhaust deposits here are an indication of engine state-of-tune. If the pipe is black and sooty or coated with white deposits, the engine is in need of a tune-up, including a thorough fuel system inspection and adjustment.

17.2   A 17mm Allen wrench or socket is required to remove the transmission fill plug

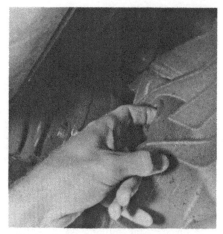

17.3   Use your finger like a dipstick to check the transmission oil level

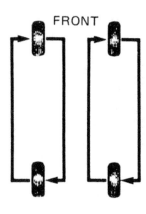

18.2   Tire rotation diagram

17   Manual transmission oil level check

Refer to illustrations 17.2 and 17.3

1   Manual transmissions do not have a dipstick. The oil level is checked by removing a plug from the side of the transmission case.
2   Remove the fill plug from the transmission case **(see illustration)**.
3   Check with your finger to make sure that the oil level is at or just below the filler hole **(see illustration)**.
4   If the transmission needs more oil, use a syringe or squeeze bottle and inject it through the plug hole.
5   Install the plug and tighten it securely. Drive the vehicle a short distance, then check for leaks.

18   Tire rotation

Refer to illustration 18.2

1   On models on which all four tires are the same size, the tires should be rotated at the specified intervals and whenever uneven wear is noticed. On some models the tires on the rear wheels are larger than those on the front wheels. On these vehicles the tires cannot be rotated. Since the vehicle will be raised and the tires removed anyway during this procedure, check the brakes (Section 19) and the wheel bearings (Section 22) at this time.
2   The tires should be rotated only in the indicated pattern **(see illustration)**.
3   Refer to the information in *Jacking and towing* at the front of this manual for the proper procedures to follow when raising the vehicle and changing a tire. If the brakes are to be checked, do not apply the parking brake as stated. Make sure the tires are blocked to prevent the vehicle from rolling.
4   Preferably, the entire vehicle should be raised at the same time. This can be done on a hoist or by jacking up each corner and then lowering the vehicle onto jackstands placed under the recommended points. Always use four jackstands and make sure the vehicle is safely supported.
5   After rotation, check and adjust the tire pressures as necessary and be sure to check the lug nut tightness.
6   For further information on the wheels and tires, refer to Chapter 10.

19   Brake system check

Refer to illustration 19.5

Note: *For detailed photographs of the brake system, refer to Chapter 9.*

Disc brakes

1   In addition to the specified intervals, the brakes should be inspected every time the wheels are removed or whenever a defect is suspected.

19.5   Check the brake pad lining thickness (arrows) by looking through the opening in the caliper

2   The brakes can be visually checked without removing any parts except the wheels.
3   Raise the vehicle and place it securely on jackstands. Remove the wheels (see *Jacking and towing* at the front of the manual, if necessary).
4   The disc brake calipers, which contain the pads, are now visible. There is an outer pad and an inner pad in each caliper. All pads should be inspected.
5   Check the pad thickness by looking through the opening in the caliper body at the pads **(see illustration)**. If the lining material is 5/64-inch or less in thickness, the pads should be replaced. Keep in mind that the lining material is riveted or bonded to a metal backing shoe and the metal portion is not included in this measurement.
6   If it is difficult to measure the exact thickness of the remaining lining material, remove the pads for further inspection or replacement (refer to Chapter 9).
7   Before installing the wheels, check for leakage and/or damage (cracks, abraded areas, etc.) around the brake hoses and connections. Replace the hose or fittings as necessary, referring to Chapter 9.
8   Check the condition of the rotor. Look for score marks, deep scratches and burned spots. If these conditions exist, the hub/rotor assembly should be removed for servicing (Chapter 9).

Parking brake

9   The easiest, and perhaps most obvious method of periodically checking the parking brake is to park the vehicle on a steep hill with

20.3 Squeeze the CV joint boot to check for cracks

22.2a On early models, pry the E-clip off the end of the speedometer cable

22.2b Dislodge each grease cap by tapping from the back with a hammer and chisel

22.3 Push on the wheel bearing thrust washer with a screwdriver (don't rest the screwdriver on the hub or locknut — it should just turn if properly adjusted)

the parking brake set and the transmission in Neutral. If the parking brake cannot prevent the vehicle from rolling, it should be adjusted (see Chapter 9).

20 Driveaxle (CV) boot check

Refer to illustration 20.3

1   The driveaxle constant velocity (CV) joints and boots should be inspected periodically or whenever the vehicle is raised.
2   Raise the vehicle and support it securely on jackstands.
3   Inspect the CV joint boots for leaks and broken retaining bands. Squeeze the boots to see if the rubber has cracked **(see illustration)**. If lubricant leaks out through a hole or crack in a boot, the CV joint will wear prematurely and require replacement. Replace any damaged boots immediately (Chapter 8).

21 Fuel system check

Warning: *There are certain precautions to take when inspecting or servicing the fuel system components. Work in a well ventilated area and do not allow open flames (cigarettes, appliance pilot lights, etc.) near*

the work area. Mop up spills immediately and do not store fuel soaked rags where they could ignite.

1   The fuel system is under pressure, so if any fuel lines are disconnected for servicing, refer to Chapter 4 for the fuel pressure relief procedure before beginning. Plug all disconnected fuel lines immediately after disconnection to prevent the tank from emptying itself.
2   The fuel system is most easily checked with the vehicle raised on a hoist so the components underneath the vehicle are readily visible and accessible.
3   If the smell of gasoline is noticed while driving or after the vehicle has been in the sun, the system should be thoroughly inspected immediately.
4   Remove the gas filler cap and check for damage, corrosion and an unbroken sealing imprint on the gasket. Replace the cap with a new one if necessary.
5   With the vehicle raised, inspect the gas tank and filler neck for punctures, cracks and other damage. The connection between the filler neck and the tank is especially critical. Sometimes a rubber filler neck will leak due to loose clamps or deteriorated rubber, problems a home mechanic can usually rectify. **Warning:** *Do not, under any circumstances, try to repair a fuel tank yourself (except rubber components).*
6   Carefully check all rubber hoses and metal lines leading away from the fuel tank. Check for loose connections, deteriorated hoses, crimped lines and other damage. Follow the lines to the front of the vehicle, carefully inspecting them all the way. Repair or replace damaged sections as necessary.
7   If a fuel odor is still evident after the inspection, refer to Section 23.

22 Front wheel bearing check, repack and adjustment

Refer to illustrations 22.2a, 22.2b, 22.3, 22.5, 22.7, 22.12, 22.17 and 22.21

Check

1   With the vehicle supported on jackstands, spin the front wheels and check for noise and excessive rolling resistance.
2   Wheel bearing adjustment on these models is checked after first removing the front wheels and the grease caps. On early models the speedometer is driven off the left front wheel, so it will be necessary to first remove the cable E-clip with a small screwdriver **(see illustration)**. Work around the outer circumference of each grease cap with a hammer and chisel to remove it **(see illustration)**.
3   To check the bearing adjustment, use a screwdriver (without supporting it on the hub or locknut) to turn the washer **(see illustration)**. If the washer turns with finger pressure only, the bearings are properly adjusted. If the washer is difficult to turn or moves easily, the bearings must adjusted, referring to the procedure in Step 21.
4   The front wheel bearings on these models should be repacked with fresh grease and adjusted as part of the routine maintenance procedure.

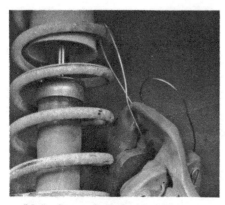

22.5   Suspend the brake caliper on a piece of wire — don't let it hang by the brake hose!

22.7   Remove the outer bearing and the thrust washer before pulling the hub off the spindle

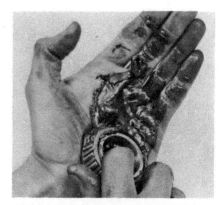

22.12   Work the grease into the bearing rollers from the back (larger side) of the race

22.17   Use a hammer and block of wood to tap the seal evenly into the hub

22.21   Adjust the locknut until the thrust washer can just be moved with the screwdriver, then tighten the Allen head clamp screw

24.2   Work the grease into the linkage contact surfaces with a brush

Repack

5   Remove the brake caliper mounting bolts, detach the caliper and hang it out of the way on a coat hanger or piece of wire **(see illustration)**.
6   Loosen the bearing locknut clamp screw with an Allen wrench, then unscrew the locknut.
7   Pull the hub out enough to dislodge the outer bearing and washer and remove them **(see illustration)**.
8   Remove the hub and place it on a workbench.
9   Use a screwdriver to carefully pry out the inner wheel bearing seal and remove the bearing.
10   Use solvent to remove all traces of old grease from the bearings, hub and spindle. A small brush may prove useful; however, make sure no bristles from the brush embed themselves inside the bearing rollers. Allow the parts to air dry.
11   Carefully inspect the bearings for cracks, score marks and worn or uneven surfaces.
12   Use an approved high temperature wheel bearing grease to pack the bearing. Work the grease completely into the bearings, forcing it between the rollers, cone and cage **(see illustration)**.
13   Apply a thin coat of grease to the spindle at the outer seat, inner bearing seat, shoulder and seal seat.
14   Put a small quantity of grease inboard of each bearing race inside the hub. Using your finger, form a dam at these points to provide extra grease availability and to keep thinned grease from flowing out of the bearing.
15   Install the grease-packed inner bearing into the rear of the hub and put a small amount of grease outboard of the bearing.
16   Lubricate the inner circumference of the new seal with a light coat of grease.
17   Place the seal over the inner bearing and tap it evenly into place using a hammer and a block of wood **(see illustration)**.
18   Carefully place the hub assembly onto the spindle and push the grease-packed outer bearing into position.
19   Install the brake caliper.

20   Install the washer and locknut and adjust the bearing preload using the following procedure.

Adjustment

21   While rotating the hub, tighten or loosen the locknut until the washer can just be moved with the screwdriver **(see illustration)**. Tighten the Allen head clamp screw securely.
22   Install the grease caps and wheels and lower the vehicle.

---

23   Evaporative emissions control system check

1   The function of the Evaporative Emissions Control System is to draw fuel vapors from the tank and fuel injection system, store them in a charcoal canister and burn them during normal engine operation.
2   The most common symptom of a fault in the evaporative emissions system is a strong fuel odor in the engine compartment. If a fuel odor is detected, inspect the charcoal canister, located in the left front wheel well.
3   The evaporative system can be checked by visually inspecting the canister and hoses in the wheel well and then tracing the hose into the engine compartment, looking for cracks and damage.
4   The evaporative emissions control system is explained in more detail in Chapter 6.

---

24   Headlight retractor linkage lubrication

Refer to illustration 24.2

1   Lubricate the contact surfaces of the headlight retractor mechanism at the specified intervals, if the linkage sticks or whenever the engine compartment has been washed.
2   A brush can be used to work white lithium base grease into the links **(see illustration)**.

Common spark plug conditions

NORMAL

Symptoms: Brown to grayish-tan color and slight electrode wear. Correct heat range for engine and operating conditions.
Recommendation: When new spark plugs are installed, replace with plugs of the same heat range.

WORN

Symptoms: Rounded electrodes with a small amount of deposits on the firing end. Normal color. Causes hard starting in damp or cold weather and poor fuel economy.
Recommendation: Plugs have been left in the engine too long. Replace with new plugs of the same heat range. Follow the recommended maintenance schedule.

CARBON DEPOSITS

Symptoms: Dry sooty deposits indicate a rich mixture or weak ignition. Causes misfiring, hard starting and hesitation.
Recommendation: Make sure the plug has the correct heat range. Check for a clogged air filter or problem in the fuel system or engine management system. Also check for ignition system problems.

ASH DEPOSITS

Symptoms: Light brown deposits encrusted on the side or center electrodes or both. Derived from oil and/or fuel additives. Excessive amounts may mask the spark, causing misfiring and hesitation during acceleration.
Recommendation: If excessive deposits accumulate over a short time or low mileage, install new valve guide seals to prevent seepage of oil into the combustion chambers. Also try changing gasoline brands.

OIL DEPOSITS

Symptoms: Oily coating caused by poor oil control. Oil is leaking past worn valve guides or piston rings into the combustion chamber. Causes hard starting, misfiring and hesitation.
Recommendation: Correct the mechanical condition with necessary repairs and install new plugs.

GAP BRIDGING

Symptoms: Combustion deposits lodge between the electrodes. Heavy deposits accumulate and bridge the electrode gap. The plug ceases to fire, resulting in a dead cylinder.
Recommendation: Locate the faulty plug and remove the deposits from between the electrodes.

TOO HOT

Symptoms: Blistered, white insulator, eroded electrode and absence of deposits. Results in shortened plug life.
Recommendation: Check for the correct plug heat range, over-advanced ignition timing, lean fuel mixture, intake manifold vacuum leaks, sticking valves and insufficient engine cooling.

PREIGNITION

Symptoms: Melted electrodes. Insulators are white, but may be dirty due to misfiring or flying debris in the combustion chamber. Can lead to engine damage.
Recommendation: Check for the correct plug heat range, over-advanced ignition timing, lean fuel mixture, insufficient engine cooling and lack of lubrication.

HIGH SPEED GLAZING

Symptoms: Insulator has yellowish, glazed appearance. Indicates that combustion chamber temperatures have risen suddenly during hard acceleration. Normal deposits melt to form a conductive coating. Causes misfiring at high speeds.
Recommendation: Install new plugs. Consider using a colder plug if driving habits warrant.

DETONATION

Symptoms: Insulators may be cracked or chipped. Improper gap setting techniques can also result in a fractured insulator tip. Can lead to piston damage.
Recommendation: Make sure the fuel anti-knock values meet engine requirements. Use care when setting the gaps on new plugs. Avoid lugging the engine.

MECHANICAL DAMAGE

Symptoms: May be caused by a foreign object in the combustion chamber or the piston striking an incorrect reach (too long) plug. Causes a dead cylinder and could result in piston damage.
Recommendation: Repair the mechanical damage. Remove the foreign object from the engine and/or install the correct reach plug.

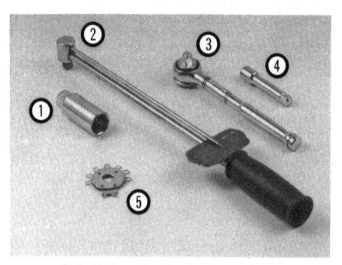

25.2   Tools required for changing spark plugs

1 **Spark plug socket** — *This will have special padding inside to protect the spark plug porcelain insulator*
2 **Torque wrench** — *Although not mandatory, use of this tool is the best way to ensure that the plugs are tightened properly*
3 **Ratchet** — *Standard hand tool to fit the plug socket*
4 **Extension** — *Depending on model and accessories, you may need special extensions and universal joints to reach one or more of the plugs*
5 **Spark plug gap gauge** — *This gauge for checking the gap comes in a variety of styles. Make sure the gap for your engine is included*

25   Spark plug replacement

Refer to illustrations 25.2, 25.5a, 24.5b, 25.9 and 25.11
Warning: *The electric cooling fan on these models can activate at any time, even when the ignition switch is in the Off position. Disconnect the fan motor or negative battery cable when working in the vicinity of the fan.*

1   The spark plugs are located on the left (driver's) side of the cylinder head.
2   In most cases the tools necessary for spark plug replacement include a plug wrench or spark plug socket which fits onto a ratchet (this special socket will be padded inside to protect the porcelain insulators on the new plugs), various extensions and a feeler gauge to check and adjust the spark plug gap **(see illustration)**. A special plug wire removal tool is available for separating the wire boot from the spark plug, but it is not absolutely necessary. Since this engine is equipped with an aluminum cylinder head, a torque wrench should be used for tightening the spark plugs.
3   The best approach when replacing the spark plugs is to purchase the new spark plugs beforehand, adjust them to the proper gap and then replace each plug one at a time. When buying the new spark plugs, be sure to obtain the correct plug for your specific engine. This information can be found on the *Emission Control Information label* located under the hood or in the factory owner's manual. If differences exist between the sources, purchase the spark plug type specified on the label as it was printed for your specific engine.
4   Allow the engine to cool completely before attempting to remove any of the plugs. During this cooling off time, each of the new spark plugs can be inspected for defects and the gaps can be checked.
5   The gap is checked by inserting the proper thickness gauge between the electrodes at the tip of the plug **(see illustration)**. The gap between the electrodes should be the same as that given in the Specifications or on the VECI label. The wire should touch each of the electrodes. If the gap is incorrect, use the adjuster on the thickness gauge body to bend the curved side electrode slightly until the proper gap is obtained **(see illustration)**. Also, at this time check for cracks in the spark plug body (if any are found, the plug should not be used). If the

25.5a   Spark plug manufacturers recommend using a wire-type gauge when checking the gap — if the wire does not slide between the electrodes with a slight drag, adjustment is required

25.5b   To change the gap, bend the *side* electrode only, as indicated by the arrows, and be very careful not to crack or chip the porcelain insulator surrounding the center electrode

side electrode is not exactly over the center one, use the adjuster to align the two.
6   Cover the front of the vehicle to prevent damage to the paint.
7   With the engine cool, remove the spark plug wire from the front spark plug. Pull only on the boot at the end of the wire; do not pull on the wire itself. Use a twisting motion to free the boot and wire from the plug. A plug wire removal tool (mentioned earlier) should be used if available.
8   If compressed air is available, use it to blow any dirt or foreign material away from the spark plug area. A common bicycle pump will also work. The idea here is to eliminate the possibility of material falling into the cylinder through the plug hole as the spark plug is removed.
9   Now place the spark plug socket over the plug and remove it from the engine by turning it in a counterclockwise direction **(see illustration)**.
10 Compare the spark plug with those shown in the accompanying color photos to get an indication of the overall running condition of the engine.
11 Apply a small amount of moly-base grease or anti-seize compound to the plug threads **(see illustration)**. Thread one of the new plugs into the hole, tightening it as much as possible by hand. **Caution:** *Be extremely careful, as these engines have aluminum cylinder heads, which means that the spark plug hole threads can be stripped easily. It may be a good idea to slip a short length of rubber hose over the end of the plug to use as a tool to thread it into place. The hose will grip the plug well enough to turn it, but will start to slip on the plug if the plug begins to cross-thread in the hole — this will prevent damaged threads and the accompanying costs involved in repairing them.*

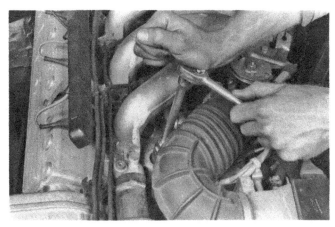

25.9   Use a socket with an extension to remove the spark plugs from the head

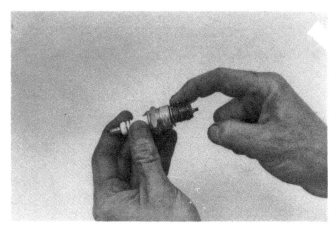

25.11   Apply moly-base grease or anti-seize compound to the threads of the new spark plugs

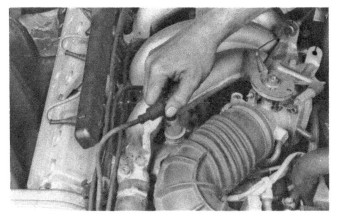

26.5   Grasp the boot (not the wire), then twist it and pull to detach the wire from the plug

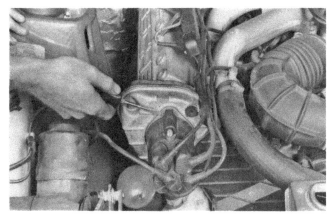

26.12   After loosening the screws, use a screwdriver to pry up the distributor cap and break the gasket seal

12 Attach the plug wire to the new spark plug, again using a twisting motion on the boot until it is seated on the spark plug.

13 Repeat the above procedure for the remaining spark plugs, replacing them one at a time to prevent mixing up the spark plug wires.

---

26   Spark plug wire, distributor cap and rotor check and replacement

---

Refer to illustrations 26.5, 26.12, 26.14, 26.15 and 26.17

Warning: *The electric cooling fan on these models can activate at any time, even when the ignition switch is in the Off position. Disconnect the fan motor or negative battery cable when working in the vicinity of the fan. The high voltage present in the ignition system can transmit a possibly fatal shock. Consequently, extreme care should be take when working near the ignition system when the engine is running.*

1   Begin this procedure by making a visual check of the spark plug wires while the engine is running. In a darkened garage (make sure there is ventilation) start the engine and observe each plug wire. Be careful not to come into contact with any moving engine parts. If there is a break in the wire, you will see arcing or a small spark at the damaged area. If arcing is noticed, make a note to obtain new wires, then allow the engine to cool and check the distributor cap and rotor.

2   Disconnect the negative cable at the battery. Place the cable out of the way so it cannot accidentally come in contact with the negative terminal of the battery, as this would once again allow power into the electrical system of the vehicle.

Spark plug wires

3   The spark plug wires should be checked at the recommended intervals and whenever new spark plugs are installed in the engine.

4   The wires should be inspected one at a time to prevent mixing up the order, which is essential for proper engine operation.

5   Disconnect the plug wire from one spark plug. To do this, grab the rubber boot, twist slightly and pull the wire free. Do not pull on the wire itself, only on the rubber boot **(see illustration)**.

6   Check inside the boot for corrosion, which will look like a white crusty powder. Push the wire and boot back onto the end of the spark plug. It should be a tight fit on the plug. If it is not, remove the wire and use a pair of pliers to carefully crimp the metal connector inside the boot until it fits securely on the end of the spark plug.

7   Using a clean rag, wipe the entire length of the wire to remove any built-up dirt and grease. Once the wire is clean, check for burns, cracks and other damage. Do not bend the wire excessively or pull the wire since the conductor inside might break.

8   Disconnect the wire from the distributor cap. Again, pull only on the rubber boot. Check for corrosion and a tight fit in the same manner as the spark plug end. Reattach the wire to the distributor cap.

9   Check the remaining spark plug wires one at a time, making sure they are securely fastened at the distributor and the spark plug when the check is complete.

10 If new spark plug wires are required, purchase a new set for your specific engine model. Wire sets are available pre-cut, with the rubber boots already installed. Remove and replace the wires one at a time to avoid mix-ups in the firing order.

Distributor cap

11 It is common practice to install a new cap and rotor whenever new spark plug wires are installed. Although the breakerless distributor used on this vehicle requires much less maintenance than conventional distributors, periodic inspections should be performed when the plug wires are inspected.

12 Loosen the two screws which hold the distributor cap to the engine. Note that these screws have catches at the ends so they do not come completely out of the cap. Carefully break the gasket seal with a small screwdriver and remove the cap **(see illustration)**.

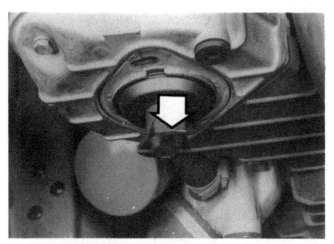

26.14   Inspect the distributor rotor for cracks and deterioration of the terminal (arrow)

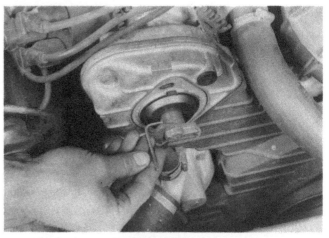

26.15   Loosen the Allen screw before removing the distributor rotor

15   To replace the rotor, first loosen the Allen screw (see illustration). The rotor is a tight fit and it may be necessary to carefully tap it off the end of the distributor shaft.
16   Install the new rotor on the distributor shaft (it is keyed to fit properly only one way) and tighten the Allen screw.
17   Before installing the distributor cap, inspect the cap for cracks and other damage. Closely examine the terminals on the inside of the cap for excessive corrosion and damage (see illustration). Slight scoring is normal. Again, if in doubt as to the condition of the cap, replace it with a new one.
18   Replace the cap and secure it with the screws.
19   Attach the spark plug wires to the cap in their proper positions.

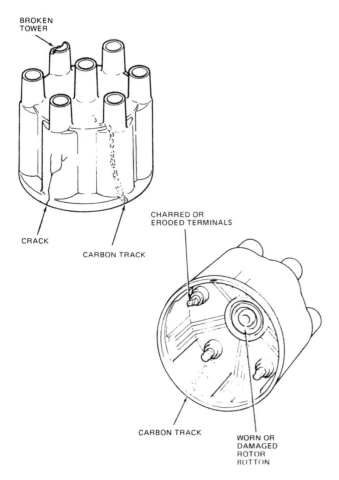

BROKEN TOWER

CRACK

CHARRED OR ERODED TERMINALS

CARBON TRACK

CARBON TRACK

WORN OR DAMAGED ROTOR BUTTON

26.17   Shown here are some of the common defects to look for when inspecting the distributor cap (if in doubt about its condition, install a new one)

13   The rotor is now visible at the center of the distributor. It is held in place by an Allen head screw.
14   Visually inspect the rotor for cracks and damage. Carefully check the condition of the metal terminal at the top of the rotor for erosion and burned areas (see illustration). If in doubt as to its condition, replace it with a new one.

27   Engine idle speed check and adjustment

The manufacturer recommends that the idle speed be checked and adjusted at the specified interval. Since this requires the use of special tools and equipment, the vehicle should be taken to a dealer or properly equipped shop to have this procedure done.

28   Clutch disc wear check

Refer to illustrations 28.3 and 28.4
1   The clutch disc wear should be checked visually at the specified intervals.
2   Raise the vehicle and support it securely on jackstands.
3   Remove the rubber inspection plug (see illustration).
4   Measure the distance between the end of the clutch slave cylinder piston and the edge of the inspection hole and compare this measurement to the Specifications (see illustration).

29   Rear axle (differential) oil (automatic transmission models) check

1   Automatic transmissions have a separate oil supply for the differential. The differential oil level is checked by removing the plug from the side of the transmission case.
2   Remove the fill plug located at the top of the transmission case.
3   Check with your finger to make sure that the oil level is at or just below the filler hole.
4   If the differential needs more oil, use a syringe to squeeze the appropriate lubricant into the opening.
5   Install the plug and tighten it securely. Drive the vehicle a short distance, then check for leaks.

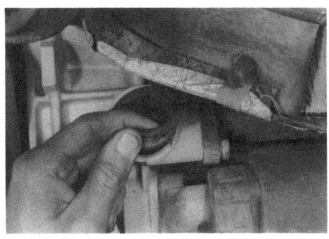

28.3   Pull the rubber plug out of the clutch disc wear inspection hole

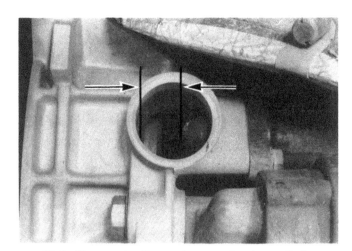

28.4   Measure the distance between the slave cylinder piston end and the inspection hole wall directly opposite it (arrows)

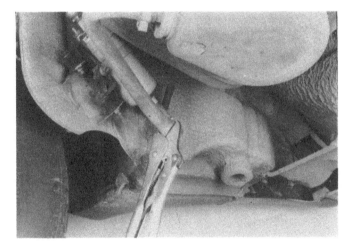

30.4   Pinch off the fuel line with vise-grip pliers that don't have teeth in the jaws (they're made especially for clamping hoses)

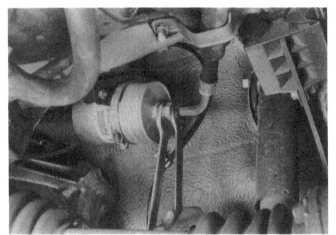

30.5   Hold the fuel filter with one wrench while unscrewing the fuel line fitting

30   Fuel filter replacement

Refer to illustrations 30.4 and 30.5

Warning: *Gasoline is extremely flammable, so extra precautions must be taken when working on any part of the fuel system. Do not smoke or allow open flames or bare light bulbs near the work area. Also, do not work in a garage if a natural gas-type appliance with a pilot light is present.*

1   The fuel filter is a disposable canister type and is located in the fuel line under the rear of the vehicle, adjacent to the fuel tank.
2   Depressurize the fuel system (Chapter 4).
3   Raise the vehicle and support it securely on jackstands.
4   Use vise-grip pliers to pinch the fuel line closed so the fuel tank won't drain while the filter is off **(see illustration)**.
5   Disconnect the fuel lines from the filter, using two wrenches (one to hold the filter and the other to unscrew the fuel line fittings) **(see illustration)**.
6   Loosen the bracket clamp screw and detach the filter from the vehicle, holding your fingers over the fittings to keep the residual fuel from running out.
7   Place the new filter in position in the bracket and tighten the clamp screw securely.
8   Connect the fuel lines to the filter. Tighten the fuel line fittings securely.
9   Remove the pliers from the fuel tank line, start the engine and check for leaks.
10   Lower the vehicle.

31   Camshaft drivebelt check and adjustment

The camshaft drivebelt must be checked and adjusted at the specified intervals. The belt could fail if it is not kept properly tensioned and cause major engine damage. Because the factory tension measurement tool necessary to perform this procedure is not available to the general public, camshaft drivebelt adjustment must be performed by a dealer service department.

32   Balance shaft drivebelt check and adjustment

The balance shaft drivebelt must be checked and adjusted at the specified intervals. Because the factory tension measurement tool necessary to perform this procedure is not available to the general public, balance shaft drivebelt adjustment must be performed by a dealer service department.

33   Cooling system servicing (draining, flushing and refilling)

Refer to illustrations 33.6a, 33.6b and 33.11

1   Periodically, the cooling system should be drained, flushed and refilled to replenish the antifreeze mixture and prevent formation of rust and corrosion, which can impair the performance of the cooling system and cause engine damage.

33.6a   Unscrew the radiator drain plug with a pair of pliers

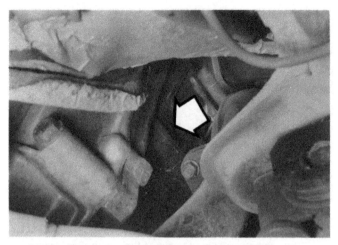

33.6b   The engine block drain plug is located under the manifold on the right side of the engine (arrow)

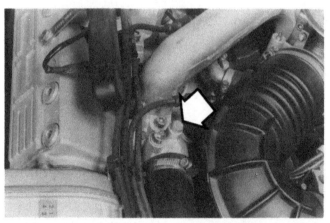

33.11   The cooling system bleed plug (arrow) is located on the upper radiator hose water outlet

2   At the same time the cooling system is serviced, all hoses and the radiator cap should be inspected and replaced if defective (see Section 7).

3   Since antifreeze is a corrosive and poisonous solution, be careful not to spill any of the coolant mixture on the vehicle's paint or your skin. If this happens, rinse immediately with plenty of clean water. Consult your local authorities about the dumping of antifreeze before draining the cooling system. In many areas, reclamation centers have been set up to collect automobile oil and drained antifreeze/water mixtures, rather than allowing them to be added to the sewage system.

4   With the engine cold and the heater set at Warm, remove the coolant reservoir cap.

5   Move a large containers under the radiator and the engine to catch the coolant as it is drained.

6   Drain the radiator and the cooling system by removing the drain plug at the bottom of the radiator and the right side of the engine block **(see illustrations)**. If either or both of the drain plugs has excessive corrosion and cannot be turned easily, disconnect the lower radiator hose to allow the coolant to drain. Be careful that none of the solution is splashed on your skin or into your eyes.

7   Disconnect the hoses from the coolant reservoir and remove the reservoir. Flush it out with clean water.

8   Install the coolant reservoir, connect the hoses, place a garden hose in the filler neck and flush the system until the water runs clear at all drain points.

9   In severe cases of contamination or clogging of the radiator, remove it (see Chapter 3) and reverse flush it. This involves inserting the hose in the bottom radiator outlet to allow the water to run against the normal flow, draining through the top. A radiator repair shop should be consulted if further cleaning or repair is necessary.

10   When the coolant is regularly drained and the system refilled with the correct antifreeze/water mixture, there should be no need to use

chemical cleaners or descalers.

11   To refill the system, reconnect the radiator hoses and install the drain plugs. Unscrew the bleed plug located at the base of the upper radiator hose until the bleed hole in the plug shaft is visible **(see illustration)**. On some models the bleed plug is located in the upper radiator hose and is held in place by a screw-type clamp.

12   Slowly fill the cooling system until the reservoir level is at the Max mark. Install the reservoir cap.

13   Run the engine until normal operating temperature is reached (the fan has turned on and off) and the coolant issuing from the bleed plug no longer has any air bubbles. Tighten the bleed plug securely. Add coolant as necessary to bring the reservoir level to between the Max and Min marks.

14   Keep a close watch on the coolant level and the cooling system hoses during the first few miles of driving. Tighten the hose clamps and/or add more coolant as necessary.

34   Automatic transmission fluid change

1   At the specified time intervals, the transmission fluid should be drained and replaced.

2   Before beginning work, purchase the specified transmission fluid (see *Recommended lubricants and fluids* at the front of this Chapter).

3   Other tools necessary for this job include jackstands to support the vehicle in a raised position, a drain pan capable of holding at least 8 pints, newspapers and clean rags.

4   The fluid should be drained immediately after the vehicle has been driven. This will remove any built up sediment better than if the fluid were cold. Because of this, it may be wise to wear protective gloves. Fluid temperature can exceed 350° in a hot transmission.

5   After the vehicle has been driven to warm up the fluid, raise it and place it on the jackstands for access underneath.

6   Move the necessary equipment under the vehicle, being careful not to touch any of the hot exhaust components.

7   Place the drain pan under the lower part of the transmission and remove the fluid tube between the pan and fluid reservoir. Be sure the drain pan is in position, as fluid will come out with some force.

8   Remove the transmission pan.

9   Remove the filter screen.

10   Clean the filter screen thoroughly with solvent.

11   Carefully remove all traces of old gasket material from the pan and transmission body (don't nick or gouge the sealing surfaces).

12   Install the filter, using a new gasket.

13   Install the pan, using a new gasket.

14   Install the fluid tube between the pan and the reservoir, making sure the fittings are tightened securely.

15   Add the specified amount and type of fluid to the reservoir (use a funnel to prevent spills). It is best to add a little fluid at a time, continually checking the level (Section 10). Allow the fluid time to drain into the pan.

36.3   A 17 mm Allen head wrench or socket is required to remove the transmission fill and drain plugs

16 With the selector lever in Park, apply the parking brake and start the engine without depressing the accelerator pedal (if possible). Do not race the engine at high speed; run at slow idle only.
17 With the engine still idling, check the level in the reservoir. Check under the vehicle for leaks.
18 Check the fluid level to make sure it is just below the Max mark. Do not allow the fluid level to go above this point, as the transmission would then be overfilled, necessitating draining of the excess fluid.
19 Drive the vehicle to reach normal operating temperature. Park the vehicle on a level surface and check the fluid level with the engine idling and the transmission in Park. The level should now be at the Max mark. If not, add more fluid to bring the level up to this point. Again, do not overfill.

35   Brake fluid replacement

1 Because brake fluid absorbs moisture which could ultimately cause corrosion of the brake components, and air which could make the braking system less effective, the fluid should be replaced at the specified intervals. This job can be accomplished for a nominal fee by a properly equipped brake shop using a pressure bleeder. The task can also be done by the home mechanic with the help of an assistant. To bleed the air and old fluid and replace it with fresh fluid from sealed containers, refer to the brake bleeding procedure in Chapter 9.
2 If there is any possibility that incorrect fluid has been used in the system, drain all the fluid and flush the system with methylated spirits. Replace all piston seals and cups, as they will be affected and could possibly fail under pressure.

36   Manual transmission oil change

Refer to illustration 36.3
Note: *Carefully read through this Section before undertaking this procedure. You will need to purchase the correct type and amount of the specified oil before draining the old oil out of the vehicle (see Recommended lubricants at the front of this chapter). You will also need a 17 mm Allen wrench to remove the drain and fill plugs.*
1 Raise the vehicle and support it securely on jackstands.
2 Move the drain pan, rags, newspapers and wrench under the transmission.
3 Remove the transmission drain and fill plugs and allow the oil to drain into the pan **(see illustration)**.
4 After the oil has drained completely, reinstall the drain plug and tighten it securely.
5 Using a hand pump, syringe or funnel, fill the transmission until the oil level is even with the bottom of the plug hole.
6 Install the plug and tighten it securely.

Chapter 2 Part A Engine

Contents

Specifications

Balance shaft diameter ....   1.2204 to 1.2210 in (30.975 to 30.991 mm)
Balance shaft bearing bore diameter ....   1.3790 to 1.3797 in (35.000 to 35.019 mm)
Balance shaft bushing bore diameter ....   1.3396 to 1.3403 in (34.000 to 34.019 mm)

Torque specifications*

	Ft-lbs	Nm
Balance shaft sprocket bolts	33	45
Balance shaft housings		
6 mm bolt	6	8
8 mm stud nut		
step 1	11	15
step 2	22	30
8 mm bolt		
step 1	11	15
step 2	15	20
Crankshaft pulley bolt	154	210
Camshaft sprocket bolt		
hex socket bolt	33	45
multiple socket bolt	48 to 52	65 to 70
Camshaft assembly-to-cylinder head bolts	15	20
Camshaft assembly threaded plugs	29	40
Cylinder head bolts	See text, Section 6	
Intake manifold bolt	15	20
Exhaust manifold nut	15	20
Engine mount nut	35	48
Chassis crossmember bolt and nut	62	85
Stabilizer bar-to-body bolt	17	23

*Additional torque specifications can be found in Chapter 2, Part B

**2479cc, 2681cc &
2969cc engines
Firing order
1-3-4-2**

Cylinder location and distributor rotation

1 General information

This Part of Chapter 2 is devoted to in-vehicle repair procedures for the engine. All information concerning engine removal and installation and engine block and cylinder head overhaul can be found in Part B of this Chapter.

Since the repair procedures included in this Part are based on the assumption that the engine is still installed in the vehicle, if they are being used during a complete engine overhaul (with the engine already out of the vehicle and on a stand) many of the steps included here will not apply.

The specifications included in this Part of Chapter 2 apply only to the procedures found here. The specifications necessary for rebuilding the block and cylinder head are included in Part B.

2 Repair operations possible with the engine in the vehicle

Many major repair operations can be accomplished without removing the engine from the vehicle.

Clean the engine compartment and the exterior of the engine with some type of pressure washer before any work is done. A clean engine will make the job easier and will help keep dirt out of the internal areas of the engine.

Depending on the components involved, it may be a good idea to remove the hood to improve access to the engine as repairs are performed (refer to Chapter 11 if necessary).

If oil or coolant leaks develop, indicating a need for gasket or seal replacement, the repairs can generally be made with the engine in the vehicle. The cylinder head gasket, intake and exhaust manifold gaskets, camshaft assembly and the balance shaft assemblies are accessible with the engine in place.

Exterior engine components, such as the water pump, the starter motor, the alternator, the distributor cap and rotor and the fuel injection components, as well as the intake and exhaust manifolds, can be removed for repair with the engine in place.

Since the cylinder head can be removed without pulling the engine, valve component servicing can also be accomplished with the engine in the vehicle.

Replacement of, repairs to or inspection of the camshaft and balance shaft belts and sprockets are all possible with the engine in place.

3 Camshaft assembly — removal and installation

Refer to illustrations 3.9, 3.10a, 3.10b, 3.12, 3.15, 3.16a and 3.16b

Removal

1 Due to the special tools and techniques necessary, disassembly of the camshaft assembly (housing, camshaft, bearings, etc.) cannot be accomplished by the home mechanic. It is possible to remove the assembly, however, so that it can be taken to a dealer or properly equipped shop.

2 Disconnect the negative cable at the battery. Place the cable out of the way so it cannot accidentally come in contact with the negative terminal of the battery, as this would once again allow power into the electrical system of the vehicle.

3 Remove the distributor cap and rotor.

4 Disconnect the throttle cable.

5 Remove the fuel injection fuel rail assembly (Chapter 4).

6 Remove the engine front cover.

7 Mark the position of the camshaft sprocket and belt.

8 Loosen the camshaft belt tensioner and pull the belt off the sprocket.

9 Unscrew the plugs in the camshaft cover with an Allen wrench for access to the upper retaining bolts **(see illustration)**.

10 Remove the camshaft assembly retaining bolts, working through the holes in the cover for access to the upper bolts **(see illustrations)**.

11 Lift the assembly up off the cylinder head while an assistant holds the valve lifters in place so they do not fall out.

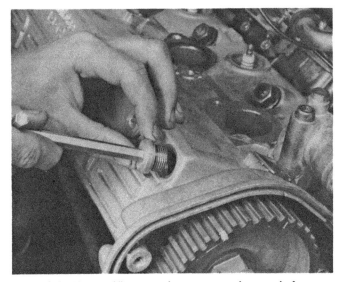

3.9 Use an Allen wrench to unscrew the camshaft assembly plugs

3.10a Remove the upper camshaft assembly bolts, working through the holes in the cover with an Allen head socket and extension

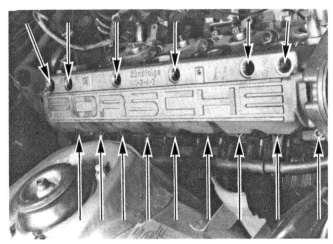

3.10b Camshaft assembly bolt locations (arrows)

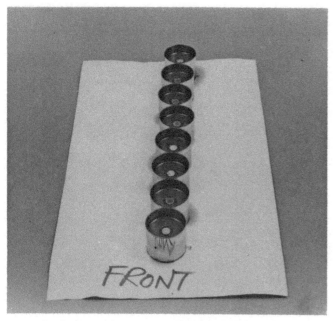

3.12   Keep the valve lifters in order so they can be
reinstalled in their original locations

3.15   The OBEN or TOP on the camshaft cover gasket
must face up

3.16a   Grease will hold the valve lifters in place while the
camshaft assembly is lowered

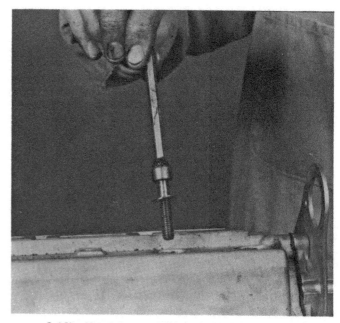

3.16b   Use grease to hold the bolts on the wrench

12 Place the camshaft assembly on a clean working surface. If the valve lifters are removed, keep them in order for installation to their original positions (see illustration).

Installation

13 The mating surfaces of the cylinder head and camshaft assembly must be perfectly clean when the camshaft assembly is installed.
14 Use a gasket scraper to remove all traces of old gasket material, taking care not to gouge the soft aluminum. Wipe the mating surfaces with a cloth saturated with lacquer thinner or acetone.
15 Place the new gasket in position on the cylinder head with the OBEN or TOP lettering facing up (see illustration).
16 Place the camshaft assembly in position using assembly lube or grease to hold the valve lifters in place (see illustration). Install the retaining bolts using grease to hold the bolts in place on the wrench (see illustration). Tighten the bolts evenly to the specified torque to draw

the camshaft assembly onto the cylinder head. Install the camshaft cover plugs and tighten to the specified torque.
17 Install the camshaft belt on the sprocket, making sure the marks made during removal line up. Caution: Any time the camshaft belt tension is released the vehicle must be taken to a dealer to have the belt retensioned using a special tool not available to the general public. Improper tensioning of the camshaft belt can cause severe engine damage.
18 The remainder of installation is the reverse of removal.

---

4   Intake manifold — removal and installation

Refer to illustrations 4.10, 4.15 and 4.17
Removal

1   Depressurize the fuel system (Chapter 4).

4.10   Cut the fuel line retaining strap with wire cutters

4.15   Remove the old gaskets with a scraper. Be careful — the head is easily damaged

4.17   Feed the fuel and vacuum lines into position before setting the intake manifold into place

2   Disconnect the negative cable at the battery. Place the cable out of the way so it cannot accidentally come in contact with the negative terminal of the battery, as this would once again allow power into the electrical system of the vehicle.
3   Remove the air intake hose from between the air flow sensor and the throttle body.
4   Disconnect the throttle cable.
5   Disconnect the vacuum hoses from the throttle body and unplug the electrical connector from the throttle position sensor.
6   Remove the oil dipstick-to-manifold bolt.
7   Remove the manifold support bracket bolts.
8   Disconnect the spark plug wires.
9   Unclip the spark plug wire harness from the fuel rail and move it out of the way.
10   Cut the fuel line retaining strap and disconnect the fuel and vacuum lines **(see illustration)**.
11   Remove the fuel rail (Chapter 4).
12   Remove the intake manifold bolts. It may be necessary to strike the intake manifold with a rubber mallet to break the gasket seal.
13   Lift the manifold up for access and disconnect the vacuum hoses.
14   Remove the manifold from the engine.

Installation

15   Using a scraper, carefully remove the old gaskets from the cylinder head **(see illustration)**.
16   Place the new intake manifold gaskets in position.
17   To install, connect the vacuum hoses on the underside, thread the vacuum and fuel hoses through the manifold and set it in position **(see illustration)**.
18   Install the bolts and tighten them to the specified torque.
19   The remaining installation steps are the reverse of removal.

---

5   Exhaust manifolds — removal and installation

Refer to illustration 5.4

1   Raise the vehicle and support it securely on jackstands.
2   In the engine compartment, remove the manifold stud nuts.
3   Under the vehicle, disconnect the exhaust system.
4   Remove manifold heat shields **(see illustration)**.
5   Lower the manifolds from the vehicle.
6   Installation is the reverse of removal.
7   Tighten the nuts in several steps until they are all at the specified torque.

5.4   The exhaust manifold heat shield is held in place by two bolts (arrows)

---

6   Cylinder head — removal and installation

Refer to illustrations 6.5, 6.6, 6.9, 6.14, 6.15 and 6.18

Removal

1   Disconnect the negative cable at the battery. Place the cable out of the way so it cannot accidentally come in contact with the negative terminal of the battery, as this would once again allow power into the electrical system of the vehicle.
2   Turn the engine over to the Top Dead Center (TDC) position for the number one piston (see Section 7).
3   Refer to Section 3 and remove the camshaft assembly.
4   Refer to Sections 4 and 5 and remove the intake and exhaust manifolds. Note that the cooling system must be drained to prevent coolant from getting into internal areas of the engine when the head is removed. Remove the thermostat (Chapter 3) to completely drain the cylinder head.

6.5   The two Allen head bolts (arrows) are located at the front corner of the cylinder head

6.6   Mark the vacuum solenoid connections with tape before disconnecting

6.9   Pry the cylinder head loose at the tab on the front corner — DO NOT pry between the gasket surfaces

6.14   Do not bend the gasket when installing it over the studs

6.15   Be careful not to contact the studs when lowering the head into place on the block

5   Remove the two Allen head bolts from the front corner of the cylinder head (see illustration).
6   Mark their positions and disconnect the vacuum switching valve hoses (see illustration).
7   Remove the water outlet housing.
8   Working from the center out, loosen and remove the cylinder head retaining nuts.
9   Lift the head up off the studs and remove it from the engine. If resistance is felt, pry carefully at the tab on the front edge to break the gasket seal (see illustration).
10   Cylinder head disassembly and inspection procedures are covered in detail in Chapter 2, Part B.

Installation

11   The mating surfaces of the cylinder head and block must be perfectly clean when the heads are installed.
12   Use a gasket scraper to remove all traces of carbon and old gasket material, then wipe the mating surfaces with a cloth saturated with lacquer thinner or acetone. If there is oil on the mating surfaces when the head is installed, the gasket may not seal correctly and leaks may develop. **Caution:** *Since the head is made of aluminum, aggressive scraping can cause damage. Be extra careful not to nick or gouge the mating surfaces with the scraper.*

13   Check the block and head mating surfaces for nicks, deep scratches and other damage. If damage is slight, it can be removed with a file. If it is excessive, machining may be the only alternative.
14   Position the new gasket over the studs (see illustration).
15   Lower the head onto the block (see illustration).
16   Before installing the head nuts, coat the threads with clean engine oil. Do not use any lubricant on the studs or washers. **Caution:** *When installing, be very careful not to drop any of the nuts or washers down into the oil galleries.*
17   Install the nuts and tighten them finger tight.
18   Tighten the nuts in three steps, in the sequence shown (see illustration).
Note: *Porsche has revised the cylinder head torque specifications for various engines and model years. Contact a Porsche dealer for the latest information regarding your specific engine type.*
19   The remainder of installation is the reverse of removal.

7   Top Dead Center (TDC) for number 1 piston — locating

Refer to illustrations 7.4 and 7.5

1   Top Dead Center (TDC) is the highest point in the cylinder that

6.18   Cylinder head nut tightening sequence

7.4   The TDC marks on the camshaft sprocket and cover (arrows) are visible after removing the rubber plug

7.5   The TDC marks on the flywheel and clutch housing (arrows) can be viewed at the rear of the engine

each piston reaches as it travels up-and-down when the crankshaft turns. Each piston reaches TDC on the compression stroke and again on the exhaust stroke, but TDC generally refers to piston position on the compression stroke. The timing marks on the flywheel, visible through the opening in the clutch housing at the rear of the engine, and the camshaft sprocket at the front of the engine, are referenced to the number one piston at TDC on the compression stroke.

2   Positioning the piston at TDC is an essential part of many procedures, such as replacement of the camshaft and balance shaft belts and sprockets.

3   In order to bring the piston to TDC, the crankshaft must be turned using one of the methods outlined below. When looking at the front of the engine, normal crankshaft rotation is clockwise. **Warning:** *Before beginning this procedure, be sure to place the transmission in Neutral and disconnect the battery negative cable.*

 a) The preferred method is to turn the crankshaft with a large socket and breaker bar attached to the pulley bolt that is threaded into the front of the crankshaft.
 b) A remote starter switch, which may save some time, can also be used. Attach the switch leads to the S (switch) and B (battery)

terminals on the starter motor. Once the piston is close to TDC, use a socket and breaker bar as described in the previous paragraph.

 c) If an assistant is available to turn the ignition switch to the Start position in short bursts, you can get the piston close to TDC without a remote starter switch. Use a socket and breaker bar to complete the procedure.

4   Use a small screwdriver to remove the rubber inspection hole plug in the camshaft cover just above the distributor cap (**see illustration**). The TDC mark on the camshaft sprocket can be viewed through this inspection hole. The number 1 piston is at TDC when the mark on the camshaft sprocket is aligned with the mark on the housing.

5   Turn the crankshaft until the line on the flywheel is aligned with the mark on the clutch housing (**see illustration**). The timing mark on the flywheel can be seen from above, looking at the rear of the engine.

6   After the number one piston has been positioned at TDC on the compression stroke, TDC for any of the remaining cylinders can be located by turning the crankshaft 180° at a time and following the firing order.

8   Balance shaft/camshaft belts and sprockets — removal and installation

Refer to illustrations 8.3, 8.4, 8.5, 8.6a, 8.6b, 8.7, 8.8a, 8.8b, 8.9, 8.10a, 8.10b, 8.12a, 8.12b, 8.13a, 8.13b, 8.14a, 8.14b, 8.15, 8.16, 8.17a, 8.17b, 8.18, 8.19, 8.20a, 8.20b, 8.21, 8.22, 8.25, 8.26 and 8.27

Caution: *Any time the tension is released from the balance shaft or camshaft drivebelts, they must be retensioned before driving the vehicle or serious engine damage can result. Since tensioning can only be performed with a special tool not available to the general public, this job must be left to a dealer. Whenever the belts, sprockets and tensioner are moved, their positions must be carefully marked so they can be returned to these original positions as described below. If great care is taken to return the belts and sprockets to their original positions, the vehicle can be driven to a dealer for final tensioning. If these procedures are not followed, it is recommended that the vehicle be towed, rather than driven, to a dealer to have the belts tensioned.*

Balance shaft belts and sprockets
Removal

1   Position the number 1 piston at TDC (Section 7).
2   Remove the alternator and (if equipped) the power steering pump drivebelt.

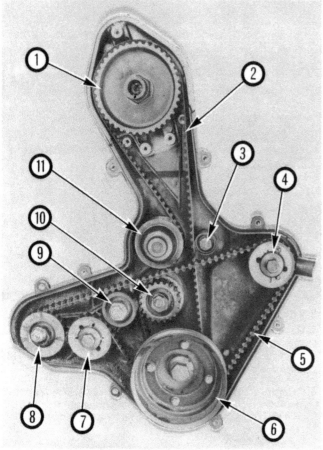

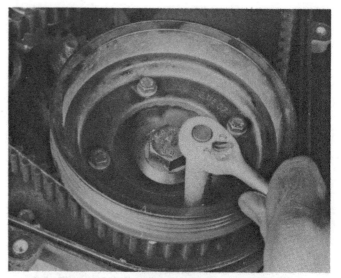

8.4   The crankshaft pulley is held in place by four bolts

8.3   Camshaft and balance shaft belt and sprocket
component layout

1 Camshaft sprocket   8 Balance shaft
2 Camshaft belt   belt tensioner
3 Camshaft belt idler   9 Balance shaft
4 Upper balance   belt idler
   shaft sprocket   10 Camshaft
5 Balance shaft belt   belt tensioner
6 Crankshaft pulley   11 Water pump pulley
7 Lower balance
   shaft sprocket

8.5   Mark the relationship of the balance shaft belt at the
crankshaft and lower balance shaft sprockets (arrows)

8.6a   Use drill bits to measure the clearance between the
lower balance shaft sprocket and adjuster . . .

8.6b   . . . and the idler (try different size bits until you find
one that just slides between them with a slight amount of drag)

8.7 To loosen the lower balance shaft tensioner, loosen the locknut and turn the adjuster nut (arrow)

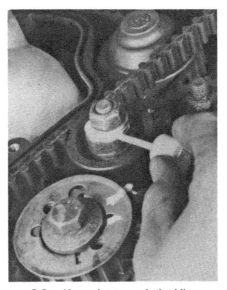

8.8a Use paint to mark the idler locknut position

8.8b Loosen the idler locknut

8.9 With the tension released, lift the belt off the lower balance shaft sprocket

8.10a Slide the tensioner . . .

3 Remove the engine front cover for access to the belts and sprockets **(see illustration)**.
4 Remove the four bolts and detach the crankshaft pulley **(see illustration)**.
5 Mark the relative position of the belt on the crankshaft sprocket and both balance shaft sprockets **(see illustration)**.
6 Measure the clearances between the lower balance shaft sprocket and tensioning sprocket and the lower balance shaft and idler with various size drill bits **(see illustrations)**. Mark these bits and set them aside so the clearances can be reset during installation.
7 Loosen the balance shaft tensioning sprocket locknut, then turn the tensioner nut counterclockwise to loosen it **(see illustration)**.
8 Mark the position of the idler with paint, then loosen the locknut to provide enough clearance to pass the belt between the idler and balance shaft sprocket **(see illustrations)**.
9 Remove the balancer belt, marking the direction of rotation and taking care not to twist it if it is going to be reused **(see illustration)**.
10 Remove the tensioner sprocket and idler by sliding them off the shafts **(see illustrations)**.

8.10b . . . and idler off their shafts

8.12a Lock each balance shaft sprocket so it won't turn while the bolt is loosened — the tool in this photo is a commonly available pin spanner

8.12b You can use two punches, locking pliers and a screwdriver to lock the balance shaft sprockets if a pin spanner isn't available

8.13a Mark the balance shaft-to-sprocket relationship with paint

8.13b Slide the balance shaft sprocket off the shaft

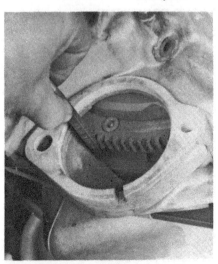

8.14a Keep the engine from turning by inserting a screwdriver or punch between the starter gear teeth and the clutch housing wall

8.14b Unscrew the crankshaft sprocket bolt

8.15 A puller is necessary to remove the crankshaft sprocket

8.16 Install the upper balance shaft sprocket with the marks (arrows) lined up

11 Mark the belt guides (flanges) and sprockets to ensure correct re-installation.
12 Remove the balance shaft sprocket bolts. It will be necessary to lock the balance shaft sprockets while removing the retaining bolts. This can be accomplished with a special tool (available at auto parts stores), which is recommended, or by using two large punches, locking pliers and a small crowbar (see illustrations).
13 Mark the relationship of each balance shaft sprocket and shaft (see illustration). Remove the sprockets by pulling them off the shafts (see illustration).
14 To remove the crankshaft sprocket bolt, first remove the starter motor and have an assistant lock the starter gear teeth with a screwdriver or punch so the crankshaft cannot turn (see illustration). With the engine locked so it won't turn, remove the sprocket bolt (see illustration).
15 Use a puller to detach the crankshaft sprocket (see illustration).

8.17a Install the balance shaft belt with the marks aligned

Installation
16 Prior to installation, make sure the balance shaft keys are facing up. Install the balance shaft sprockets. Make sure the notches in the sprockets line up with the pointers on the cover and install the belt guides (see illustration).
17 Install the tensioner sprocket, idler and crankshaft sprocket, followed by the belt. Carefully return all of the sprockets and idlers to their marked positions and tighten the tensioning sprocket bolts and locknuts securely (see illustrations). Remember that the belt must be tensioned by a dealer even if all components are returned to their original positions. This is particularly true with a new belt because it will stretch when the engine is run.

Camshaft belt and sprockets
Removal
18 Remove the balance shaft drivebelt. If the camshaft belt is to be reused, mark the direction of rotation and the relationship of the belt and crankshaft sprocket (see illustration).

8.17b Using the drill bit, return the balancer shaft tensioner sprocket to its original position, then tighten the locknut while holding the sprocket in place

8.18 Mark the camshaft belt direction of rotation and relationship to the crankshaft sprocket if the belt is to be reused

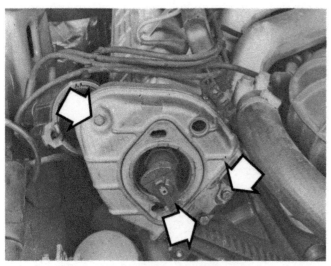

8.19   The upper camshaft cover is held in place by three bolts (arrows)

8.20a   Use a drill bit to measure the clearance between the water pump pulley and the camshaft belt tensioning sprocket — try different size bits until you find one that just slides between them with a slight amount of drag

8.20b   Mark the position of the lower edge of the tensioner with paint (arrow) so the adjustment can be restored when the belt is reinstalled

8.21   Use white paint to mark the relative positions of the camshaft sprocket and belt

19 Remove the distributor cap, followed by the upper camshaft sprocket cover (see illustration).
20 Measure the clearance between the water pump pulley and the tensioning sprocket with a suitable drill bit so this clearance can be attained when the belt is reinstalled (see illustration). Mark the position of the tensioning sprocket with paint to make installation easier (see illustration).
21 Mark the relative position of the belt and the camshaft pulley (see illustration).
22 Loosen the locknut and turn the belt tension adjusting nut beneath it clockwise to remove the tension from the belt (see illustration).
23 Remove the belt, taking care not to twist it if it is going to be re-installed.
24 Remove the bolts and slide the tensioning sprocket and idler off the shafts.
25 Detach the sprocket using a suitable puller (see illustration).

Installation
26 Make sure the camshaft sprocket is at aligned with the TDC marks on the housing and install the sprockets and idler (see illustration). Also check to be sure the number 1 piston is still at TDC.
27 Install the belt on the crankshaft sprocket, thread it over the tensioning sprocket and the water pump pulley and then onto the camshaft sprocket (see illustration). Return the tensioning sprocket to the marked position and tighten it securely.
28 The remainder of installation is the reverse of removal. Again,

8.22   Loosen the camshaft belt tensioner locknut and turn the tensioner clockwise to release the belt

8.25 Use a puller to remove the crankshaft sprocket

8.26 The TDC marks on the camshaft and housing (arrows) must be lined up before installing the belt (this picture was taken with the belt already installed)

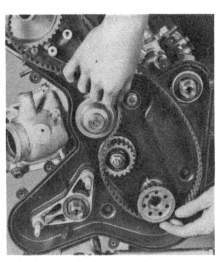

8.27 Loop the camshaft belt over the crankshaft sprocket and then thread it over the other sprockets

9.3 Pry the keys out of the balance shafts with a small screwdriver

9.4 Use a plastic mallet to dislodge the balance shaft front support housing from the engine

remember that a seriously out-of-adjustment camshaft belt can allow the valves to strike the pistons, causing severe engine damage. Consequently the belt must be taken to a dealer for checking and adjustment any time the tension has been released.

9 Balance shafts — removal and installation

Refer to illustrations 9.3, 9.4, 9.7, 9.8, 9.9, 9.10, 9.11, 9.12, 9.15a, 9.15b, 9.15c, 9.16a, 9.16b, 9.17a, 9.17b, 9.18, 9.19, 9.21, 9.22, 9.24a, 9.24b and 9.24c

Removal

1   Remove the engine front cover.
2   Remove the balance shaft belts and sprockets (Section 8).
3   Use a screwdriver to pry out the balance shaft keys **(see illustration)**.
4   Remove the bolts and remove the front balance shaft support housings. It may be necessary to use a plastic mallet to dislodge the housings **(see illustration)**.
5   Remove the bolts and nuts, pry carefully on the tabs at the corners to dislodge the housings and lift the housings off.
6   Lift out the balance shafts, taking care not to contact the bearing surfaces on the studs.

Installation

7   Remove the end rings from the housings **(see illustration)**.

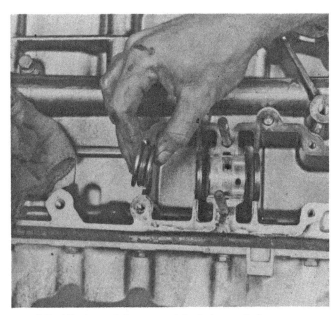

9.7 Lift the end rings out of the balance shaft grooves

9.8   Pry the O-ring off the end ring

9.9   Insert a screwdriver under the inner edge of the seal and pry it out

9.10   Lift the spacer out of the housing

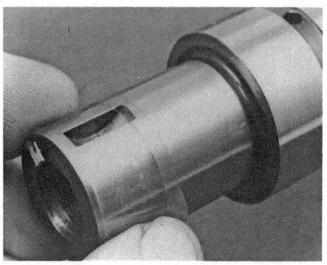

9.11   The front balance shaft seal is very thin. Slide it on the shaft very carefully

9.12   Coat the contact surface of the spacer with grease or assembly lube

9.15a   The seals are marked with arrows showing the direction of shaft rotation. The left seal goes in the lower balance shaft support and the right one in the upper support

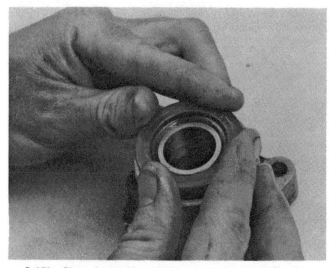

9.15b   Place the seal into the front of the support housing

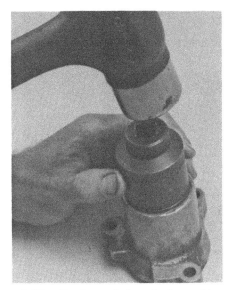

9.15c Tap it evenly into the bore with a socket

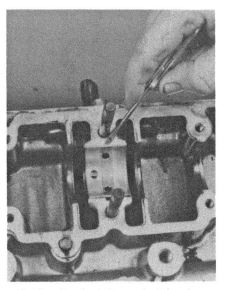

9.16a Pry the balance shaft bearing out with a small screwdriver inserted into the groove in the housing

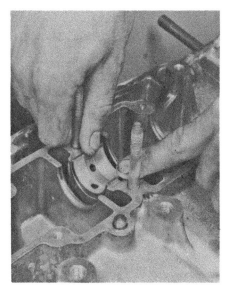

9.16b Press the new bearing into place

9.17a Coat the bearings . . .

9.17b . . . and balance shaft contact surfaces with assembly lube

9.18 Insert the end ring into the housing groove

8   Remove the old O-ring from the end ring and install a new one **(see illustration)**.
9   Pry the old oil seal out of the front support housing **(see illustration)**.
10 Remove the spacer from the front support housing **(see illustration)**.
11 Carefully slide the very thin seal onto the end of each balance shaft, taking care to avoid damaging it **(see illustration)**.
12 Coat the spacer contact surface with assembly lube or moly base grease **(see illustration)**.
13 Install the spacer into the housing.
14 Coat the lip of each seal with clean oil.
15 Note that the seals are designed for opposite directions of rotation and are marked with arrows **(see illustration)**. Carefully install the seals over the spacer in the appropriate support housing and tap them into place with a suitable socket and hammer **(see illustrations)**.
16 Inspect the bearings for damage or scoring, replacing with new ones if necessary **(see illustrations)**.
17 Apply a thin, uniform layer of clean moly-base grease or engine assembly lube to the contact surfaces of the shaft and bearings **(see illustrations)**.
18 Lubricate the end ring with clean engine oil and install it into the groove in the block **(see illustration)**.
19 Remove the O-ring from the shaft front support housing **(see illustration)**.

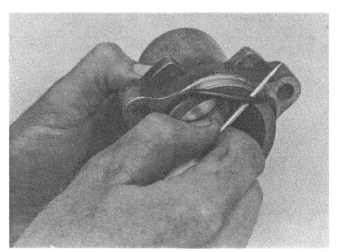

9.19 Remove the balance shaft front support O-ring with a screwdriver

9.21 Lower the balance shaft into place while pushing the
front support housing onto the end

9.22 Apply the sealant evenly to the balance shaft
housing contact surface

9.24a Lower balance shaft bolt and nut
tightening sequence

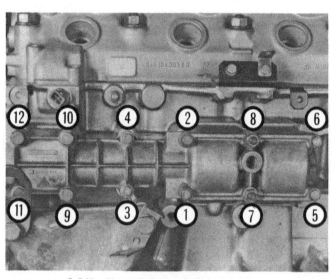

9.24b Upper balance shaft bolt and nut
tightening sequence

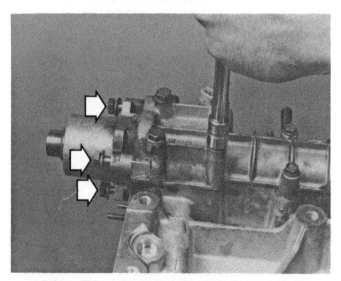

9.24c Tighten the front support bolts (arrows) after
tightening the housing bolts

20 Install a new O-ring in the groove and lubricate it with clean engine
oil.
21 Lower the balance shaft into position and slide the front support
assembly onto the end if the shaft (see illustration).
22 Apply a thin coat of Loctite 638 sealant or equivalent to the shaft
housing contact surface (see illustration).
23 Place the housing in position on the block.
24 Install the bolts and nuts finger tight. Tighten the housing-to-block
bolts in the specified sequence, in two steps, to the specified torque
(see illustrations). Tighten the front support bolts to the specified torque
(see illustration).
25 The remainder of installation is the reverse of removal.

---

10 Engine mounts — check and replacement

Refer to illustration 10.12
Check

1   The hydraulic engine mounts used on these models seldom require
attention. Symptoms of a failed engine mount are knocking or vibration
when the engine is started or stopped. Check the mounts visually for
deterioration or separation of the rubber from the metal and for leaking fluid.

9.25 Balance shaft component layout

1 Flat washer
2 Housing bolt
3 Upper shaft sprocket bolt
4 Sprocket collar (belt guide)
5 Sprocket
6 Shaft seal
7 Spacer
8 Seal
9 Lower bearing housing
10 O-ring
11 Washer
12 Lower shaft woodruff key
13 Upper shaft woodruff key
14 Lower balance shaft
15 Bearing shell
16 Seal
17 End ring
18 Bearing shell
19 Lower shaft sprocket bolt
20 Washer
21 Sprocket collar (belt guide)
22 Sprocket
23 Shaft seal
24 Spacer
25 Seal
26 Bolt
27 Washer
28 O-ring
29 Upper balance shaft housing
30 Washer
31 Bolt
32 Washer
33 Bolt
34 Washer
35 Bolt
36 Washer
37 Stud nut

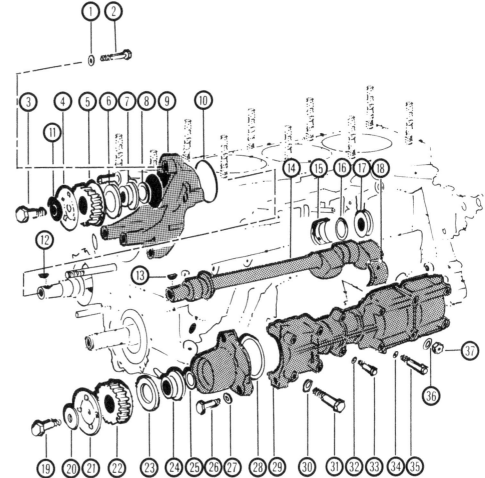

0033-H

Replacement

2   Disconnect the negative cable at the battery. Place the cable out of the way so it cannot accidentally come in contact with the negative terminal of the battery, as this would once again allow power into the electrical system of the vehicle.

3   Raise the vehicle and support it on jackstands. Connect a hoist to the engine and raise it sufficiently to remove the engine weight from the mounts.

4   Remove the under engine splash guard (if equipped).

5   Remove the stabilizer bar (Chapter 10).

6   Remove the heat shield from the right side engine mount.

7   Remove the engine support-to-mount nuts and bolts.

8   On power steering equipped models, disconnect and plug the fluid hose at the pump.

9   On later models, unclip the brake pad wear sensor wire.

10 Mark the steering shaft universal joint-to-shaft relationship.

11 Remove the mount-to-crossmember nuts. Remove the stabilizer bar support-to-body bolts.

12 Remove the front crossmember-to-chassis bolts (see illustration).

13 Pull the front crossmember down sufficiently to provide clearance and disconnect the steering shaft universal joint.

14 Rotate the top of the each mount out of the crossmember and remove them from the vehicle toward the front, turning each 180° if necessary.

15 To install, insert each engine mount securely into position in the crossmember.

16 Raise the crossmember with a jack while guiding the steering shaft into the marked position in the universal joint.

17 Continue raising the crossmember until the tops of the engine mounts are secure in the engine supports. Install the crossmember bolts.

18 Install the engine support-to-mount nuts and bolts.

19 Install the steering shaft universal bolt.

20 Tighten the engine mount-to-crossmember nuts to the specified torque.

10.12 Remove the crossmember-to-chassis bolts (arrows) and lower the crossmember with a jack so the mounts can be lowered from the vehicle

21 Tighten the engine mount-to-support bolts to the specified torque.

22 Tighten the crossmember bolts to the specified torque.

23 Install any components which were removed during the replacement procedure.

Chapter 2   Part B
General engine overhaul procedures

Contents

Specifications

General

Bore and stroke
  (2.5 L) .... 3.94 in (100 mm) x 3.11 in (78.9 mm)
  (2.7 L) .... 4.09 in (104 mm) x 3.11 in (78.9 mm)
  (3.0 L) .... 4.09 in (104 mm) x 3.46 in (88 mm)
Compression pressure .... 145 psi (10 bar) or more

Engine block

Out-of-round limit .... 0.0007 in (0.020 mm)

Pistons and rings

Piston-to-cylinder bore clearance
  standard .... 0.0003 to 0.0012 in (0.008 to 0.032 mm)
  service limit .... 0.0031 in (0.080 mm)
Piston ring-to-groove side clearance
  Mahle pistons
    top groove .... 0.0019 to 0.0032 in (0.05 to 0.082 mm)
    second groove .... 0.0015 to 0.0028 in (0.04 to 0.072 mm)
    third groove .... 0.0009 to 0.0053 in (0.023 to 0.137 mm)
  KS pistons
    top groove .... 0.0019 to 0.0032 in (0.05 to 0.082 mm)
    second groove .... 0.0019 to 0.0032 in (0.05 to 0.082 mm)
    third groove .... 0.0009 to 0.0053 in (0.023 to 0.137 mm)
Piston ring end gap
  standard
    top compression ring .... 0.007 to 0.017 in (0.20 to 0.45 mm)
    second compression ring .... 0.007 to 0.017 in (0.20 to 0.45 mm)
    oil control ring .... 0.015 to 0.055 in (0.38 to 1.40 mm)
Piston pin
  diameter .... 0.945 in (23.996 mm)
  pin-to-piston clearance limit .... 0.0007 to 0.0012 in (0.018 to 0.032 mm)

Crankshaft and flywheel

Main journal diameter . . . .   2.7568 to 2.7576 in (69.971 to 69.990 mm)
Thrust bearing width limit . . . .   1.185 in
Main bearing oil clearance
    standard . . . .   0.0007 to 0.0038 in (0.020 to 0.098 mm)
    wear limit . . . .   0.0063 in (0.16 mm)
Connecting rod journal diameter . . . .   2.047 to 2.048 in (51.971 to 51.990 mm)
Connecting rod bearing oil clearance
    standard . . . .   0.0007 to 0.0027 in (0.02 to 0.07 mm)
    wear limit . . . .   0.0039 in (0.10 mm)
Crankshaft end play . . . .   0.0043 to 0.0122 in (0.110 to 0.312 mm)
Balance shaft diameter . . . .   1.2204 to 1.2210 in (30.975 to 30.991 mm)
Balance shaft bearing bore diameter . . . .   1.3790 to 1.3797 in (35.000 to 35.019 mm)
Balance shaft bushing bore diameter . . . .   1.3396 to 1.3403 in (34.000 to 34.019 mm)

Cylinder head and valve train

Head warpage limit . . . .   0.003 in (0.08 mm)
Valve seat angle . . . .   45°
Valve length
    intake . . . .   4.393 in (111.5 mm)
    exhaust . . . .   4.408 in (111.9 mm)
Valve seat width
    intake . . . .   0.0669 in (1.7 mm)
    exhaust . . . .   0.0788 in (2.0 mm)
Valve head diameter
    intake . . . .   1.77 in (45.0 mm)
    exhaust . . . .   1.57 in (40.0 mm)
Valve face angle . . . .   45°
Minimum valve margin width . . . .   1/32 in
Valve stem-to-guide clearance . . . .   0.031 in (0.8 mm) maximum
Valve guide inside diameter . . . .   0.3446 to 0.3552 in (9.000 to 9.015 mm)
    wear limit . . . .   0.031 in (0.80 mm)
Valve stem diameter
    intake . . . .   0.3529 to 0.3534 in (8.958 to 8.970 mm)
    exhaust . . . .   0.3521 to 0.3526 in (8.938 to 8.950 mm)

Torque specifications*

	Ft-lbs	Nm
Alternator mount bolt	33	45
Balance shaft sprocket bolt	33	45
Balance shafts		
6 mm bolt	6	8
8 mm stud nut		
step 1	11	15
step 2	22	30
8 mm bolt		
step 1	11	15
step 2	15	20
Central tube-to-clutch housing bolt	31	42
Clutch cover-to-engine bolt	55	75
Connecting rod nut		
smooth surface nut	46	62
ribbed surface nut	59	80
Crankshaft pulley bolt	154	210
Crankcase lower-to-upper section		
12 mm nuts		
step 1	15	20
step 2	29	40
step 3	55	75
10 mm nuts		
step 1	15	20
step 2	37	50
8 mm nuts	15	20
6 mm nuts	7	10
Camshaft sprocket bolt		
hex socket bolt	33	45
multiple socket bolt	48 to 52	65 to 70
Camshaft assembly-to-cylinder head bolts	15	20
Camshaft assembly threaded plugs	29	40
Cylinder head nuts	See text, Section 6, Part A	
Engine front cover bolt	6	8
Intake manifold bolt	15	20

Torque specifications (continued)*

	Ft-lbs	Nm
Engine speed sensor bracket bolt	15	20
Exhaust manifold nut	15	20
Flywheel/driveplate-to-crankshaft bolt	66	90
Oil pressure relief valve plug	33	45
Oil pressure sensor	26	35
Oil pump bolt		
6 mm	6	8
10 mm	33	45
Engine mount nut	35	48
Oil pan bolts		
step 1	3	4
step 2	7	9.5
Spark plugs	18 to 22	25 to 30
Water pump bolt	6	8

*Additional torque specifications can be found in Chapter 2, Part A

1 General information

Included in this portion of Chapter 2 are the general overhaul procedures for the cylinder head and internal engine components. The information ranges from advice concerning preparation for an overhaul and the purchase of replacement parts to detailed, step-by-step procedures covering removal and installation of internal engine components and the inspection of parts.

The following Sections have been written based on the assumption that the engine has been removed from the vehicle. For information concerning in-vehicle engine repair, as well as removal and installation of the external components necessary for the overhaul, see Part A of this Chapter and Section 7 of this Part.

The specifications included here in Part B are only those necessary for the inspection and overhaul procedures which follow. Refer to Part A for additional specifications.

2 Engine removal — methods and precautions

If you have decided that an engine must be removed for overhaul or major repair work, several preliminary steps should be taken.

Locating a suitable work area is extremely important. A shop is, of course, the most desirable place to work. Adequate work space, along with storage space for the vehicle, will be needed. If a shop or garage is not available, at the very least a flat, level, clean work surface made of concrete or asphalt is required.

Cleaning the engine compartment and engine before beginning the removal procedure will help keep tools clean and organized.

An engine hoist or A-frame will also be necessary. Make sure that the equipment is rated in excess of the combined weight of the engine and its accessories. Safety is of primary importance, considering the potential hazards involved in removing the engine from the vehicle.

If the engine is being removed by a novice, a helper should be available. Advice and aid from someone more experienced would also be helpful. There are many instances when one person cannot simultaneously perform all of the operations required when removing the engine from the vehicle.

Plan the operation ahead of time. Arrange for or obtain all of the tools and equipment you will need prior to beginning the job. Some of the equipment necessary to perform engine removal and installation safely and with relative ease are (in addition to an engine hoist) a heavy duty floor jack, complete sets of wrenches and sockets as described in the front of this manual, wooden blocks and plenty of rags and cleaning solvent for mopping up spilled oil, coolant and gasoline. If the hoist is to be rented, make sure that you arrange for it in advance and perform beforehand all of the operations possible without it. This will save you money and time.

Plan for the vehicle to be out of use for a considerable amount of time. A machine shop will be required to perform some of the work which the do-it-yourselfer cannot accomplish due to a lack of special equipment. These shops often have a busy schedule, so it would be

wise to consult them before removing the engine in order to accurately estimate the amount of time required to rebuild or repair components that may need work.

Always use extreme caution when removing and installing the engine. Serious injury can result from careless actions. Plan ahead, take your time and a job of this nature, although major, can be accomplished successfully.

3 Engine overhaul — general information

It is not always easy to determine when, or if, an engine should be completely overhauled, as a number of factors must be considered.

High mileage is not necessarily an indication that an overhaul is needed, while low mileage does not preclude the need for an overhaul. Frequency of servicing is probably the most important consideration. An engine that has had regular and frequent oil and filter changes, as well as other required maintenance, will most likely give many thousands of miles of reliable service. Conversely, a neglected engine may require an overhaul very early in its life.

Excessive oil consumption is an indication that piston rings and/or valve guides are in need of attention. Make sure that oil leaks are not responsible before deciding that the rings and/or guides are bad. Have a cylinder compression or leakdown test performed by an experienced tune-up mechanic to determine the extent of the work required.

If the engine is making obvious knocking or rumbling noises, the connecting rod and/or main bearings are probably at fault.

Loss of power, rough running, excessive valve train noise and high fuel consumption rates may also point to the need for an overhaul, especially if they are all present at the same time. If a complete tune-up does not remedy the situation, major mechanical work is the only solution.

An engine overhaul involves restoring the internal parts to the specifications of a new engine. During an overhaul, the piston rings are replaced and the cylinder walls are reconditioned. If a rebore is done, new pistons are required. Note that because of the special materials and coatings used on the block and cylinder walls, boring and honing can only be done by a dealer service department or a shop with the necessary specialized equipment. The main bearings and connecting rod bearings are generally replaced with new ones and, if necessary, the crankshaft may be reground to restore the journals. Generally, the valves are serviced as well, since they are usually in less-than-perfect condition at this point. While the engine is being overhauled, other components, such as the starter and alternator, can be rebuilt as well. The end result should be a like new engine that will give many trouble free miles.

Before beginning the engine overhaul, read through the entire procedure to familiarize yourself with the scope and requirements of the job. Overhauling an engine is not difficult, but it is time consuming. Plan on the vehicle being tied up for a minimum of two weeks, especially if parts must be taken to an automotive machine shop for repair or reconditioning. Check on availability of parts and make sure that any necessary special tools and equipment are obtained in advance. Most work can be done with typical hand tools, although a number of precision measuring tools are required for inspecting parts to determine if they must be replaced. Often an automotive machine shop will handle the inspection of parts and offer advice concerning reconditioning and replacement. **Note:** *Always wait until the engine has been completely disassembled and all components, especially the engine block, have been inspected before deciding what service and repair operations must be performed by an automotive machine shop.*

Since the block's condition will be the major factor to consider when determining whether to overhaul the original engine or buy a rebuilt one, never purchase parts or have machine work done on other components until the block has been thoroughly inspected. As a general rule, time is the primary cost of an overhaul, so it does not pay to install worn or substandard parts.

As a final note, to ensure maximum life and minimum trouble from a rebuilt engine, everything must be assembled with care in a spotlessly clean environment.

4 Cylinder compression check

1 A compression check will tell you what mechanical condition the upper end (pistons, rings, valves, head gasket) of your engine is in. Specifically, it can tell you if the compression is down due to leakage caused by worn piston rings, defective valves and seats or a blown head gasket. **Note:** *The engine must be at normal operating temperature for this check and the battery must be fully charged.*

2 Begin by cleaning the area around the spark plugs before you remove them (compressed air works best for this). This will prevent dirt from getting into the cylinders as the compression check is being done. Remove all of the spark plugs from the engine.

3 Block the throttle wide open and disconnect the primary wires from the coil.

4 With the compression gauge in the number one spark plug hole, crank the engine over at least four compression strokes and watch the gauge. The compression should build up quickly in a healthy engine. Low compression on the first stroke, followed by gradually increasing pressure on successive strokes, indicates worn piston rings. A low compression reading on the first stroke, which does not build up during successive strokes, indicates leaking valves or a blown head gasket (a cracked head could also be the cause). Record the highest gauge reading obtained.

5 Repeat the procedure for the remaining cylinders and compare the results to the Specifications.

6 Add some engine oil (about three squirts from a plunger-type oil can) to each cylinder, through the spark plug hole, and repeat the test.

7 If the compression increases after the oil is added, the piston rings are definitely worn. If the compression does not increase significantly, the leakage is occurring at the valves or head gasket. Leakage past the valves may be caused by burned valve seats and/or faces or warped, cracked or bent valves.

8 If two adjacent cylinders have equally low compression, there is a strong possibility that the head gasket between them is blown. The appearance of coolant in the combustion chambers or the crankcase would verify this condition.

9 If the compression is unusually high, the combustion chambers are probably coated with carbon deposits. If that is the case, the cylinder head should be removed and decarbonized.

10 If compression is way down or varies greatly between cylinders, it would be a good idea to have a leak-down test performed by an automotive repair shop. This test will pinpoint exactly where the leakage is occurring and how severe it is.

5 Engine rebuilding alternatives

The do-it-yourselfer is faced with a number of options when performing an engine overhaul. The decision to replace the engine block, piston/connecting rod assemblies and crankshaft depends on a number of factors, with the number one consideration being the condition of the block. Other considerations are cost, access to machine shop facilities, parts availability, time required to complete the project and the extent of prior mechanical experience on the part of the do-it-yourselfer.

Some of the rebuilding alternatives include:

Individual parts — If the inspection procedures reveal that the engine block and most engine components are in reusable condition, purchasing individual parts may be the most economical alternative. The block, crankshaft and piston/connecting rod assemblies should all be inspected carefully.

Crankshaft kit — This rebuild package consists of a reground crankshaft and a matched set of pistons and connecting rods. The pistons will already be installed on the connecting rods. Piston rings and the necessary bearings will be included in the kit.

Short block — A short block consists of an engine block with a crankshaft and piston/connecting rod assemblies already installed. All new bearings are incorporated and all clearances will be correct. The existing camshaft assembly, valve train components, cylinder head and external parts can be bolted to the short block with little or no machine shop work necessary.

Long block — A long block consists of a short block plus an oil pump, oil pan, cylinder head, camshaft assembly and valve train components, timing and balance sprockets and belts and front cover. All components are installed with new bearings, seals and gaskets incorporated throughout. The installation of manifolds and external parts is all that is necessary.

Give careful thought to which alternative is best for you and discuss the situation with local automotive machine shops, auto parts dealers or parts store countermen before ordering or purchasing replacement parts.

6 Engine — removal and installation

Refer to illustrations 6.5, 6.7, 6.11, 6.12, 6.17, 6.18, 6.26, 6.31, 6.32, 6.36 and 6.39

Removal

1 The engine is removed and installed from underneath on these models. Raise the vehicle and support it securely on jackstand so there is at least 21 inches clearance.

2 Remove the front wheels.

3 Cover the fenders with pads or blankets.

4 Disconnect the battery ground cable, followed by the positive cable. Push the positive cable through the opening in the firewall into the engine compartment.

5 Disconnect the DEE control unit plug in the passenger compartment (Chapter 4) and pull it through the opening in the firewall into the engine compartment **(see illustration)**.

6 Drain the coolant and remove the radiator.

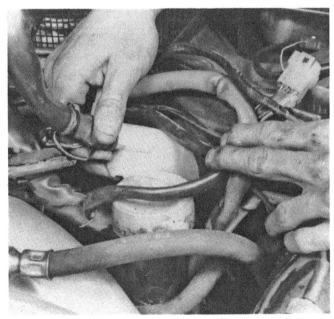

6.5 Unplug the DEE control unit plug and guide it into the engine compartment

6.7 After disconnecting the wiring harness plugs, cut the retaining strap

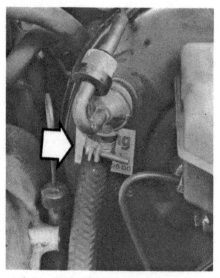

6.11 Disconnect the brake booster clamp and hose (arrow)

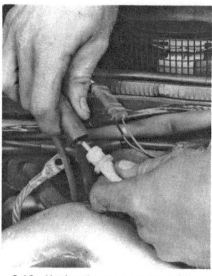

6.12 Unplug the starter motor wiring connector at the rear of the engine

6.17 Loosen the heater hose clamp and disconnect the hose from the pipe

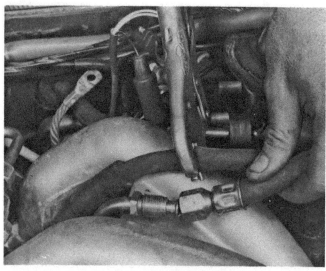

6.18 Use locking pliers to clamp the fuel return hose before disconnecting the fuel lines

6.26 Remove the stabilizer bar bolts (arrows) at the chassis

6.31 After marking their relationship, drive off the steering shaft universal joint at the stub with a hammer and punch

6.32 Disconnect the suspension arm from the chassis by removing the bolts (arrows)

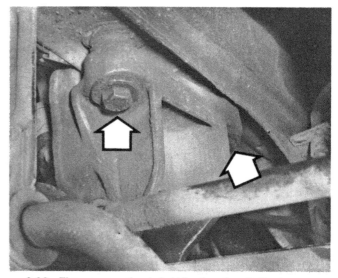

6.36 The crossmember is held in place four bolts, two on each side (arrows)

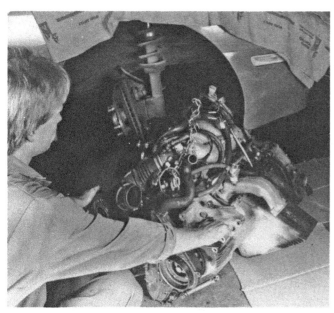

6.39 Slide the engine out from under the vehicle on a piece of carpet or cardboard

7 Unplug the wiring harness connectors at the firewall and cut the retaining straps so the harness can be removed with the engine (see illustration).
8 Disconnect the throttle cable.
9 Disconnect the engine ground cable at the firewall.
10 Remove the air cleaner housing and bracket and the air flow sensor assembly.
11 Loosen the clamp and disconnect the hose from the brake booster (see illustration).
12 Unplug the starter wiring connector in the engine compartment (see illustration).
13 Remove the starter.
14 Disconnect the power steering hoses and remove the pump.
15 Remove the clutch slave cylinder bracket bolt.
16 Mark the hoses with tape for ease of reinstallation and disconnect any vacuum hoses from the engine.
17 Disconnect the heater hose at the rear of the cylinder head (see illustration).
18 Pinch the fuel return hose shut with locking pliers and disconnect the inlet and return fuel lines (see illustration).
19 Check to make sure there are no hoses, connectors or cables which

would interfere with engine removal before proceeding.
20 Support the engine with a suitable hoist.
21 Remove the under engine splash shield.
22 Remove the exhaust system.
23 Unbolt the clutch slave cylinder and move it out of the way, leaving the fluid line connected.
24 Remove the right side engine mount shield.
25 Remove the alternator.
26 Remove the front stabilizer bar assembly-to-chassis bolts (see illustration).
27 On air conditioning equipped models, remove the compressor and hang it out of the way on a piece of wire from the spring strut with the refrigerant hoses intact.
28 Remove the lower radiator hose.
29 Remove the central tube-to-clutch housing bolts.
30 Disconnect the steering tie rod ends.
31 Mark the steering shaft and universal joint relationship and disconnect the steering gear shaft using a punch and hammer (see illustration).
32 Remove the suspension arm-to-chassis bolts (see illustration).
33 Disconnect the balljoints.
34 In the engine compartment, remove the engine mount bolts and nuts.
35 Support the crossmember with a jack, using a block of wood to protect the surface.
36 With the engine supported from above with the hoist and the crossmember/engine assembly supported from below with the jack, remove the crossmember bolts (see illustration).
37 Detach the crossmember complete with stabilizer bar and steering gear, and lower it from the vehicle.
38 Lower the engine down and forward, out of the engine compartment.
39 Carefully lower the engine onto a piece of carpeting or cardboard, detach the hoist and slide the engine out from under the vehicle (see illustration).
40 After reattaching the hoist to the engine, move the engine to a clean working surface.

Installation

41 Prior to installation, lubricate the driveshaft input splines with moly base grease.
42 With the engine blocked in an upright position on the piece of carpeting or cardboard, slide it under the vehicle and center it below the engine compartment.
43 Connect the engine hoist and raise the engine into place. Use a jack under the clutch housing to raise the rear of the engine, align the driveshaft with the splines and move the engine rearward until the driveshaft is seated in the clutch splines.
44 Install the drive tube bolts and tighten them evenly to pull the drive-

shaft splines into the crankshaft pilot bearing. Tighten the bolts to the specified torque.

45   Raise the crossmember into position with the jack and insert the bolts through the balljoints.

46   Connect the balljoints.

47   Connect the steering shaft universal joint and install the bolts. Tighten the bolts securely.

48   Raise the lower control arms into place with a jack and install the bolts.

49   Connect the steering tie rod ends.

50   Install the sway bar.

51   Install the engine mounts.

52   Install the air conditioning compressor (if equipped).

53   Install the starter.

54   Install the clutch slave cylinder.

55   Install the radiator.

56   The remainder of installation is the reverse of removal.

---

7   Engine overhaul — disassembly sequence

---

1   It is much easier to disassemble and work on the engine if it is mounted on a portable engine stand. These stands can often be rented quite cheaply from an equipment rental yard. Before the engine is mounted on a stand, the flywheel/driveplate should be removed from the engine (refer to Chapter 8).

2   If a stand is not available, it is possible to disassemble the engine with it blocked up on a sturdy workbench or on the floor. Be extra careful not to tip or drop the engine when working without a stand.

3   If you are going to obtain a rebuilt engine, all external components must come off first, to be transferred to the replacement engine, just as they will if you are doing a complete engine overhaul yourself. These include:

Alternator and brackets
Distributor cap, plug wires and spark plugs
Thermostat
Engine front cover
Camshaft and balance shaft belts and sprockets
Water pump
Fuel injection components
Intake/exhaust manifolds
Oil filler housing
Oil filter
Oil cooler assembly
Engine mounts
Clutch and flywheel/driveplate

Note: *When removing the external components from the engine, pay close attention to details that may be helpful or important during installation. Note the installed position of gaskets, seals, spacers, pins, washers, bolts and other small items.*

4   If you are obtaining a short block, which consists of the engine block, crankshaft, pistons and connecting rods all assembled, then the cylinder head, oil pan and oil pump will have to be removed as well. See *Engine rebuilding alternatives* for additional information regarding the different possibilities to be considered.

5   If you are planning a complete overhaul, the engine must be disassembled and the internal components removed in the following order:

Crankshaft pulley
Front cover
Camshaft and balance shaft belts and sprockets
Camshaft assembly
Exhaust and intake manifolds
Balance shaft assemblies
Cylinder head
Water pump
Oil filter
Oil cooler
Oil filter housing
Oil pan
Oil pump
Oil pickup
Piston/connecting rod assemblies
Crankshaft and main bearings

6   Before beginning the disassembly and overhaul procedures, make sure the following items are available:

Common hand tools
Small cardboard boxes or plastic bags for storing parts
Gasket scraper
Metric Allen head wrench set
Micrometers
Telescoping gauges
Dial indicator set
Valve spring compressor
Piston ring groove cleaning tool
Electric drill motor
Tap and die set
Wire brushes
Oil gallery brushes
Cleaning solvent

---

8   Cylinder head — disassembly

Refer to illustrations 8.3a and 8.3b

Note: *New cylinder heads are commonly available at dealerships. Due to the fact that some specialized tools are necessary for the disassembly and inspection procedures, and replacement parts may not be readily available, it may be more practical and economical for the home mechanic have the head reconditioned rather than taking the time to disassemble, inspect and recondition the original.*

1   Cylinder head disassembly involves removal of the intake and exhaust valves and related components. If it is still in place, remove the camshaft assembly (Chapter 2, Part A).

2   Before the valves are removed, arrange to label and store them, along with their related components, so they can be kept separate and reinstalled in the same valve guides they are removed from.

3   Compress the springs on the first valve with a spring compressor and remove the keepers **(see illustration)**. Carefully release the valve spring compressor and remove the retainer, the shield, the springs and the spring seat. Remove the seal from the upper end of the valve stem **(see illustration)**. If the valve binds in the guide (won't pull through), push it back into the head and deburr the area around the keeper groove with a fine file or whetstone.

4   Repeat the procedure for the remaining valves. Remember to keep all the parts for each valve together so they can be reinstalled in the same locations.

5   Once the valves and related components have been removed and stored in an organized manner, the head should be thoroughly cleaned and inspected. If a complete engine overhaul is being done, finish the engine disassembly procedures before beginning the cylinder head cleaning and inspection process.

---

9   Cylinder head — cleaning and inspection

Refer to illustrations 9.11, 9.13, 9.15 and 9.17

1   Thorough cleaning of the cylinder head and related valve train components, followed by a detailed inspection, will enable you to decide how much valve service work must be done during the engine overhaul.

Cleaning

2   Scrape away all traces of old gasket material and sealing compound from the head gasket, intake manifold and exhaust manifold sealing surfaces. Be very careful not to gouge the cylinder head. Special gasket removal solvents, which soften gaskets and make removal much easier, are available at auto parts stores.

3   Remove any built up scale from the coolant passages.

4   Run a stiff wire brush through the oil holes to remove any deposits that may have formed in them.

5   Run an appropriate size tap into each of the threaded holes to remove any corrosion and thread sealant that may be present. If compressed air is available, use it to clear the holes of debris produced by this operation.

6   Clean the exhaust and intake manifold stud threads with an appropriate size die.

7   Clean the cylinder head with solvent and dry it thoroughly. Com-

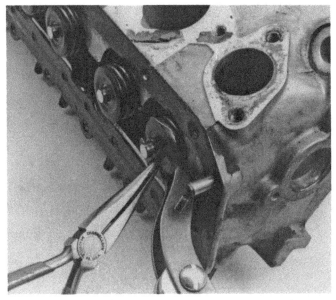

8.3a Use a valve spring compressor to compress the spring, then remove the keepers from the valve stem

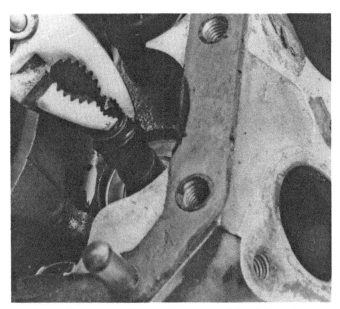

8.3b Grasp the seal with pliers and pull it off the valve

9.11 Check the cylinder head gasket surface for warpage by trying to slip a feeler gauge under the straightedge

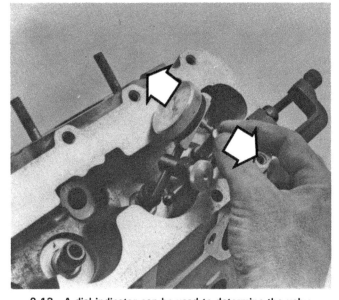

9.13 A dial indicator can be used to determine the valve stem-to-guide clearance (move the valve stem as indicated by the arrows)

pressed air will speed the drying process and ensure that all holes and recessed areas are clean. **Note:** *Decarbonizing chemicals are available and may prove very useful when cleaning cylinder heads and valve train components. They are very caustic and should be used with caution. Be sure to follow the instructions on the container.*

8   Clean all the valve springs, shields, keepers and retainers with solvent and dry them thoroughly. Do the components from one valve at a time to avoid mixing up the parts.

9   Scrape off any heavy deposits that may have formed on the valves, then use a motorized wire brush to remove deposits from the valve heads and stems. Again, make sure the valves do not get mixed up.

Inspection

Cylinder head

10 Inspect the head very carefully for cracks, evidence of coolant leakage and other damage. If cracks are found, a new cylinder head should be obtained.

11 Using a straightedge and feeler gauge, check the head gasket mating surface for warpage **(see illustration)**. If the warpage exceeds

the specified limit, it can be resurfaced at an automotive machine shop.

12 Examine the valve seats in each of the combustion chambers. If they are pitted, cracked or burned, the head will require valve service that is beyond the scope of the home mechanic.

13 Check the valve stem-to-guide clearance by measuring the lateral movement of the valve stem with a dial indicator attached securely to the head **(see illustration)**. The valve must be in the guide and approximately 1/16-inch off the seat. The total valve stem movement indicated by the gauge needle must be divided by two to obtain the actual clearance. After this is done, if there is still some doubt regarding the condition of the valve guides they should be checked by an automotive machine shop (the cost should be minimal).

Valves

14 Carefully inspect each valve face for uneven wear, deformation, cracks, pits and burned spots. Check the valve stem for scuffing and galling and the neck for cracks. Rotate the valve and check for any obvious indication that it is bent. Look for pits and excessive wear on the end of the stem. The presence of any of these conditions indicates the need for valve service by an automotive machine shop.

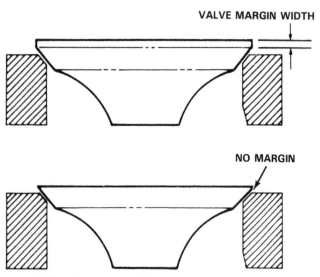

9.15   The margin width on each valve must be as specified. If no margin exists, the valve cannot be reused

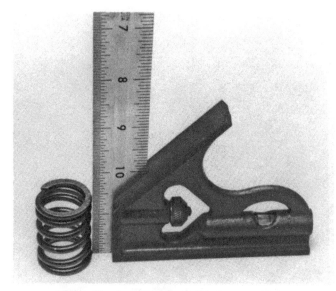

9.17   Check each valve spring for squareness

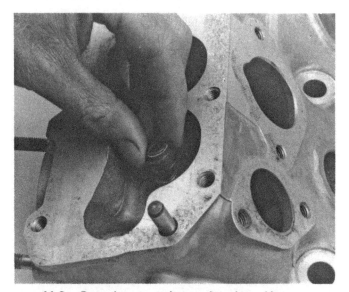

11.3a   Press the new seals over the valve guides . . .

11.3b   . . . and tap them securely into place with a socket and hammer

15   Measure the margin width on each valve (see illustration). Any valve with a margin narrower than 1/32-inch will have to be replaced with a new one.

Valve components

16   Check each valve spring for wear (on the ends) and pits. The tension of all springs should be checked with a special fixture before deciding that they are suitable for use in a rebuilt engine (take the springs to an automotive machine shop for this check).
17   Stand each spring on a flat surface and check it for squareness (see illustration). If any of the springs are distorted or sagged, replace all of them with new parts.
18   Check the spring retainers and keepers for obvious wear and cracks. Any questionable parts should be replaced with new ones, as extensive damage will occur if they fail during engine operation.
19   If the inspection process indicates that the valve components are in generally poor condition and worn beyond the limits specified, which is usually the case in an engine that is being overhauled, reassemble the valves in the cylinder head and refer to Section 10 for valve servicing recommendations.
20   If the inspection turns up no excessively worn parts, and if the valve faces and seats are in good condition, the valve train components can be reinstalled in the cylinder head without major servicing. Refer to the appropriate Section for the cylinder head reassembly procedure.

10   Valves — servicing

1   Because of the complex nature of the job and the special tools and equipment needed, servicing of the valves, the valve seats and the valve guides, commonly known as a valve job, is best left to a professional.
2   The home mechanic can remove and disassemble the head, do the initial cleaning and inspection, then reassemble and deliver the head to a dealer service department or an automotive machine shop for the actual valve servicing.
3   The dealer service department, or automotive machine shop, will remove the valves and springs, recondition or replace the valves and valve seats, recondition the valve guides, check and replace the valve springs, spring retainers and keepers (as necessary), replace the valve seals with new ones, reassemble the valve components and make sure the installed spring height (this will have to be checked by a dealer) is correct. The cylinder head gasket surface will also be resurfaced if it is warped.
4   After the valve job has been performed by a professional, the head will be in like new condition. When the head is returned, be sure to clean it again before installation on the engine to remove any metal particles and abrasive grit that may still be present from the valve ser-

11.5a   Install the valve shim
(if equipped) . . .

11.5b   . . . spring seat . . .

11.5c   . . . spring . . .

11.5d   . . . and retainer

11.6   With the spring compressed, insert the keepers.
Note the grease on the keeper, which will help hold
it in place

vice or head resurfacing operations. Use compressed air, if available, to blow out all the oil holes and passages.

11   Cylinder head — reassembly

Refer to illustrations 11.3a, 11.3b, 11.5a, 11.5b, 11.5c, 11.5d and 11.6

1   Regardless of whether or not the head was sent to an automotive repair shop for valve servicing, make sure it is clean before beginning reassembly.
2   If the head was sent out for valve servicing, the valves and related components will already be in place.
3   Install new seals on each of the valve guides. Using a hammer and a deep socket, gently tap each seal into place until it is completely seated on the guide (see illustrations). Do not twist or cock the seals

during installation or they will not seal properly on the valve stems.
4   Beginning at one end of the head, lubricate and install the first valve and apply moly-base grease or clean engine oil to the valve stem.
5   Drop the shim and spring seat over the valve guide and set the valve springs and retainer in place (see illustrations).
6   Apply a small dab of grease to each keeper to hold it in place if necessary and compress the springs with a valve spring compressor. Position the keepers in the upper groove, then slowly release the compressor and make sure the keepers seat properly (see illustration).
7   Repeat the procedure for the remaining valves. Be sure to return the components to their original locations — do not mix them up!
8   Have the installed valve spring height checked by a dealer — this requires a special tool not available to the general public. If the heads were sent out for service work, the installed height should be correct (but don't automatically assume that it is). If the height is greater than specified, shims can be added under the springs to correct it. **Caution:** *The springs should not be shimmed to the point where the installed height is less than specified.*

12.1   A socket with a universal joint is necessary to reach the oil pan bolts

12.4   Line up the metal inserts in the gasket with the bolt holes

12.5   Guide the oil pan over the pickup and then lower it onto the block

12.6   Use a punch to align the oil pan and block bolt holes

12.7   Oil pan bolt tightening sequence

13.2   Grasp the key securely with pliers and pull it up and out of the groove

13.5a   Pry carefully at the tab to dislodge the pump

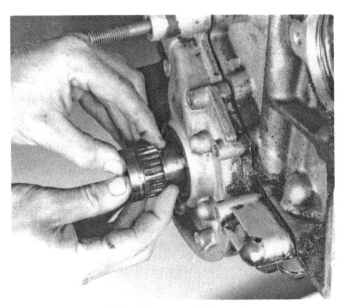

13.5b   Slide gear off the shaft

13.7   Dislodge the seal by working around the circumference, tapping with a hammer and punch

12   Oil pan — removal and installation

Refer to illustrations 12.1, 12.4, 12.5, 12.6 and 12.7

Removal

1   Remove the oil pan mounting bolts **(see illustration)**.
2   Carefully separate the pan from the block. Lift the pan off, while guiding the oil pickup out of the pan cavity.

Installation

3   Wipe the sealing surfaces of the pan and block with a cloth saturated with lacquer thinner or acetone. Make sure the bolt holes in the block are clean.
4   Place the new rubber oil pan gasket in place on the block, making sure all of the bolt holes line up **(see illustration)**.
5   Place the oil pan in position, guiding the oil pickup into the cavity in the bottom of the pan **(see illustration)**.
6   Install the bolts, using a punch or similar tool to center them **(see illustration)**.

7   Tighten the bolts to the specified torque in two steps in the sequence shown **(see illustration)**.

13   Oil pump — removal

Refer to illustrations 13.2, 13.5a, 13.5b and 13.7

1   Remove the oil pan as described in Section 12.
2   Remove the key from the pump shaft with a pair of pliers **(see illustration)**.
3   Remove the washer from the shaft.
4   Remove the mounting bolts.
5   Pry the oil pump away from the block at the tab. Dislodge the gear from the housing and slide the gear off the shaft **(see illustrations)**.
6   Lift the pump off the engine.
7   Use a punch and hammer to drive the old seal out of the back of the pump **(see illustration)**.

14.1   The oil cooler cover is held in place by two bolts

14.2   Remove the cooler housing bolts

14.3   Be careful not to drop the cooler element when withdrawing it

14.4   Pull the gasket slowly out of the groove so that it will come out in one piece

14.5   Remove the relief spring and plunger from the block

14.6a   Remove the sealing ring . . .

14.6b   . . . and the O-ring from the housing

15.4a   To prevent damage to the crankshaft journals and cylinder walls, slip sections of hose over the rod bolts before removing the pistons

15.4b   Push on the bearing insert with a hammer . . .

15.4c   . . . while supporting the piston as it comes out of the bore

14   Oil cooler — removal

Refer to illustrations 14.1, 14.2, 14.3, 14.4, 14.5, 14.6a and 14.6b

1   Remove the cooler housing cover **(see illustration)**.
2   Remove the bolts and detach the housing from the engine **(see illustration)**.
3   Lift the cooler from the engine block, taking care not to drop it **(see illustration)**.
4   Peel the gasket off the block **(see illustration)**.
5   Withdraw the relief valve and plunger from the block **(see illustration)**.
6   Remove the sealing and O-rings **(see illustrations)**.

15   Piston/connecting rod assembly — removal

Refer to illustrations 15.4a, 15.4b and 15.4c

Note: *Prior to removing the piston/connecting rod assemblies, remove the cylinder head, the oil pan and the oil pump by referring to the appropriate Sections.*

1   Turn the engine upside-down so the crankshaft is facing up.
2   Check the connecting rods and caps for identification marks. If they are not plainly marked, use a small center punch to make the appropriate number of indentations on each rod and cap (1 — 4, depending on the cylinder they are associated with).
3   Loosen each of the connecting rod cap nuts 1/2-turn at a time until they can be removed by hand.
4   Remove the number one connecting rod cap and bearing insert. Do not drop the bearing insert out of the cap. Slip a short length of plastic or rubber hose over each connecting rod cap bolt to protect the crankshaft journal and cylinder wall when the piston is removed **(see illustration)**. Push the connecting rod/piston assembly out through the top of the engine. Use a wooden hammer handle to push on the upper bearing insert in the connecting rod while guiding the piston out of the bore **(see illustrations)**.
5   Note that, unlike conventional engines, removal of the wear ridge at the top of the cylinder will not be necessary before removing the pistons. Because of a special treatment of the cylinder walls, no wear ridge should be present. Should the engine be worn to the point where there is a ridge preventing piston removal, the removal should be left to a dealer service department or shop with the special tools and equipment necessary for cylinder wall treatment.
6   Repeat the procedure for the remaining cylinders. After removal, reassemble the connecting rod caps and bearing inserts in their respec-

tive connecting rods and install the cap nuts finger tight. Leaving the old bearing inserts in place until reassembly will help prevent the connecting rod bearing surfaces from being accidentally nicked or gouged.

16   Crankshaft — removal

Refer to illustrations 16.1, 16.5 and 16.6

Note: *The crankshaft can be removed only after the engine has been removed from the vehicle. It is assumed that the flywheel or driveplate, vibration damper, camshaft and balance shaft belts and sprockets, oil pan, oil pickup, oil pump and piston/connecting rod assemblies have already been removed.*

1   Before the crankshaft is removed, check the end play. Mount a dial indicator with the stem in line with the crankshaft and just touching one the end of the crankshaft **(see illustration)**.

16.1   Checking crankshaft end play with a dial indicator

16.5   Dislodge the lower crankcase section by prying at the tab while protecting block surface with a folded cloth

16.6   Grasp the crankshaft securely when lifting it out. It is very heavy

2   Push the crankshaft all the way to the rear and zero the dial indicator. Next, pry the crankshaft to the front as far as possible and check the reading on the dial indicator. The distance that it moves is the end play. If it is greater than specified, check the crankshaft thrust surfaces for wear. If no wear is evident, new main bearings should correct the end play.

3   If a dial indicator is not available, feeler gauges can be used. Gently pry or push the crankshaft all the way to the front of the engine. Slip feeler gauges between the crankshaft and the front face of the thrust main bearing to determine the clearance.

4   Loosen each of the lower crankcase section nuts and bolts 1/4-turn at a time each, until they can be removed by hand.

5   Pry carefully at the tab located at the front corner to dislodge the lower crankcase section **(see illustration)**. Lift the lower crankcase section off. Try not to drop the bearing inserts if they come out with the section.

6   Carefully lift the crankshaft out of the engine **(see illustration)**. It is a good idea to have an assistant available, since the crankshaft is quite heavy. Slide the one-piece bearing off the front of the crankshaft. Keep the bearing inserts in place in the engine block and lower crankcase section.

17   Engine block — cleaning

1   Using a gasket scraper, remove all traces of gasket material from the engine block. Be very careful not to nick or gouge the gasket sealing surfaces.

2   Remove the main bearing bearing inserts from the upper and lower sections of the engine block. Tag the bearings, indicating which cylinder they were removed from and whether they were in the upper or lower section, then set them aside.

3   If the engine is extremely dirty it should be taken to an automotive machine shop to be steam cleaned.

4   After the block is returned, clean all oil holes and oil galleries one more time. Brushes specifically designed for this purpose are available at most auto parts stores. Flush the passages with warm water until the water runs clear, dry the block thoroughly and wipe all machined surfaces with a light, corrosion preventative oil. If you have access to compressed air, use it to speed the drying process and to blow out all the oil holes and galleries.

5   If the block is not extremely dirty or sludged up, you can do an adequate cleaning job with warm soapy water and a stiff brush. Take plenty of time and do a thorough job. Regardless of the cleaning method used, be sure to clean all oil holes and galleries very thoroughly, dry the block completely and coat all machined surfaces with light oil.

6   The threaded holes in the block must be clean to ensure accurate

torque readings during reassembly. Run the proper size tap into each of the holes to remove any corrosion, thread sealant or sludge and to restore any damaged threads. If possible, use compressed air to clear the holes of debris produced by this operation.

7   If the engine is not going to be reassembled right away, cover it with a large plastic trash bag to keep it clean.

18   Engine block — inspection

Refer to illustrations 18.4a, 18.4b and 18.4c

1   Before the block is inspected, it should be cleaned as described in Section 17.

2   Visually check the block for cracks and corrosion. Look for stripped threads in the threaded holes. It is also a good idea to have the block checked for hidden cracks by an automotive machine shop that has the special equipment to do this type of work. If defects are found, have the block repaired, if possible, or replaced.

3   Check the cylinder bores for scuffing and scoring.

4   Measure the diameter of each cylinder at the top, center and bottom of the cylinder bore, parallel to the crankshaft axis **(see illustrations)**. Next, measure each cylinder's diameter at the same three locations across the crankshaft axis. Compare the results to the Specifications. If the cylinder walls are badly scuffed or scored, or if they are out-of-round or tapered beyond the limits given in the Specifications, the block will have to be serviced by a dealer service department or a shop with the special tools and equipment necessary to replace the special cylinder wall treatment. **Caution:** *The cylinder walls of this block cannot be honed or bored with conventional boring or honing tools.*

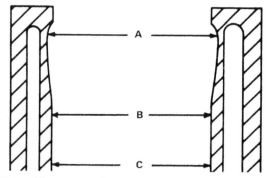

18.4a   Measure the diameter of each cylinder just under the wear ridge (A), at the center (B) and at the bottom (C)

18.4b The ability to feel when the telescoping gauge is at the correct point will be developed over time, so work slowly and repeat the check until you are satisfied that the bore measurement is accurate

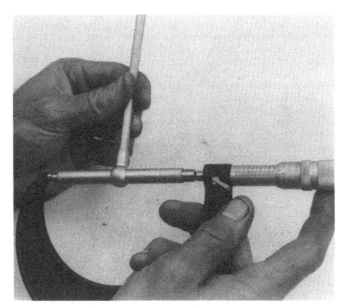

18.4c The gauge is measured with a micrometer to determine the bore size

5   If the cylinders are in reasonably good condition and not worn to the outside of the limits, and if the piston-to-cylinder clearances can be maintained properly, then they do not have to be rebored or honed (Section 19).

19   Cylinder honing

1   Prior to engine reassembly, the cylinder bores may have to be honed so the new piston rings will seat correctly and provide the best possible combustion chamber seal. Because the cylinder block on these models is made of a special aluminum alloy, honing requires special equipment and techniques. Consequently this job must be left to a dealer or an automotive machine shop.

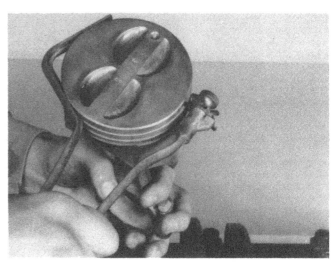

20.4a The piston ring grooves can be cleaned with a special tool, as shown here, . . .

20   Piston/connecting rod assembly — inspection

Refer to illustrations 20.4a, 20.4b, 20.10, 20.11, 20.12a, 20.12b and 20.12c

1   Before the inspection process can be carried out, the piston/connecting rod assemblies must be cleaned and the original piston rings removed from the pistons. **Note:** *Always use new piston rings when the engine is reassembled.*
2   Using a piston ring installation tool, carefully remove the rings from the pistons. Be careful not to nick or gouge the pistons in the process.
3   Scrape all traces of carbon from the top of the piston. A hand-held wire brush or a piece of fine emery cloth can be used once the majority of the deposits have been scraped away. Do not, under any circumstances, use a wire brush mounted in a drill motor to remove deposits from the pistons. The piston material is soft and will be eroded away by the wire brush.
4   Use a piston ring groove cleaning tool to remove carbon deposits from the ring grooves. If a tool is not available, a piece broken off the old ring will do the job. Be very careful to remove only the carbon deposits — don't remove any metal and do not nick or scratch the sides of the ring grooves **(see illustrations)**.
5   Once the deposits have been removed, clean the piston/rod assemblies with solvent and dry them with compressed air (if available). Make sure that the oil return holes in the back sides of the ring grooves are clear.

20.4b   . . . or a section of a broken ring

6   If the pistons are not damaged or worn excessively, and if the engine block is not rebored, new pistons will not be necessary. Normal piston wear appears as even vertical wear on the piston thrust surfaces and slight looseness of the top ring in its groove. New piston rings, on the other hand, should always be used when an engine is rebuilt.

7   Carefully inspect each piston for cracks around the skirt, at the pin bosses and at the ring lands.

8   Look for scoring and scuffing on the thrust faces of the skirt, holes in the piston crown and burned areas at the edge of the crown. If the skirt is scored or scuffed, the engine may have been suffering from overheating and/or abnormal combustion, which caused excessively high operating temperatures. The cooling and lubrication systems should be checked thoroughly. A hole in the piston crown is an indication that abnormal combustion (preignition) was occurring. Burned areas at the edge of the piston crown are usually evidence of spark knock (detonation). If any of the above problems exist, the causes must be corrected or the damage will occur again.

9   Corrosion of the piston, in the form of small pits, indicates that coolant is leaking into the combustion chamber and/or the crankcase. Again, the cause must be corrected or the problem may persist in the rebuilt engine.

10   Measure the piston ring side clearance by laying a new piston ring in each ring groove and slipping a feeler gauge between the ring and the edge of the ring groove **(see illustration)**. Check the clearance at three or four locations around each groove. Be sure to use the correct ring for each groove; they are different. If the side clearance is greater than specified, new pistons will have to be used.

11   Check the piston-to-bore clearance by measuring the bore (see Section 18) and the piston diameter. Make sure that the pistons and bores are correctly matched. Measure the piston across the skirt, at a 90° angle to and in line with the piston pin **(see illustration)**. Subtract the piston diameter from the bore diameter to obtain the clearance. If it is greater than specified, the block will have to be rebored and new pistons and rings installed.

12   Check the piston-to-rod clearance by twisting the piston and rod in opposite directions. Any noticeable play indicates that there is excessive wear, which must be corrected. The piston can be separated from the rod by removing the snap-ring with a screwdriver and pushing the pin out with a suitable socket and extension **(see illustrations)**. Inspect the pin bushing and pin for damage and wear. Measure the pin and compare the measurements to Specifications **(see illustration)**.

13   If worn or damaged, the piston/connecting rod assemblies should be taken to an automotive machine shop to have the pistons and rods rebored and new pins installed. While they are there have the connecting rods checked for bend and twist, since automotive machine shops have special equipment for this purpose.

14   Check the connecting rods for cracks and other damage. Temporarily remove the rod caps, lift out the old bearing inserts, wipe the rod and cap bearing surfaces clean and inspect them for nicks, gouges and scratches. After checking the rods, replace the old bearings, slip the caps into place and tighten the nuts finger tight.

21   Crankshaft – inspection

Refer to illustration 21.2

1   Clean the crankshaft with solvent and dry it with compressed air (if available). Be sure to clean the oil holes with a stiff brush and flush them with solvent. Check the main and connecting rod bearing journals for uneven wear, scoring, pits and cracks. Check the rest of the crankshaft for cracks and other damage.

2   Using a micrometer, measure the diameter of the main and connecting rod journals and compare the results to the Specifications **(see illustration)**. By measuring the diameter at a number of points around each journal's circumference, you will be able to determine whether or not the journal is out-of-round. Take the measurement at each end of the journal, near the crank throws, to determine if the journal is tapered.

3   If the crankshaft journals are damaged, tapered, out-of-round or worn beyond the limits given in the Specifications, have the crankshaft reground by an automotive machine shop. Be sure to use the correct size bearing inserts if the crankshaft is reconditioned.

4   Refer to Section 22 and examine the main and rod bearing inserts.

22   Main and connecting rod bearings – inspection

1   Even though the main and connecting rod bearings should be replaced with new ones during the engine overhaul, the old bearings should be retained for close examination, as they may reveal valuable information about the condition of the engine.

2   Bearing failure occurs because of lack of lubrication, the presence of dirt or other foreign particles, overloading the engine and corrosion. Regardless of the cause of bearing failure, it must be corrected before the engine is reassembled to prevent it from happening again.

3   When examining the bearings, remove them from the engine block, the lower crankcase section, the connecting rods and the rod caps and lay them out on a clean surface in the same general position as their location in the engine. This will enable you to match any bearing problems with the corresponding crankshaft journal.

4   Dirt and other foreign particles get into the engine in a variety of ways. If may be left in the engine during assembly, or it may pass through filters or the oil system. It may get into the oil, and from there into the bearings. Metal chips from machining operations and normal engine wear are often present. Abrasives are sometimes left in engine components after reconditioning, especially when parts are not thoroughly cleaned using the proper cleaning methods. Whatever the source, these foreign objects often end up embedded in the soft bearing material and are easily recognized. Large particles will not embed in the bearing and will score or gouge the bearing and journal. The best prevention for this cause of bearing failure is to clean all parts thoroughly and keep everything spotlessly clean during engine assembly. Frequent and regular engine oil and filter changes are also recommended.

5   Lack of lubrication (or lubrication breakdown) has a number of interrelated causes. Excessive heat (which thins the oil), overloading (which squeezes the oil from the bearing face) and oil leakage or throw off (from excessive bearing clearances, worn oil pump or high engine speeds) all contribute to lubrication breakdown. Blocked oil passages, which usually are the result of misaligned oil holes in a bearing shell, will also oil starve a bearing and destroy it. When lack of lubrication is the cause of bearing failure, the bearing material is wiped or extruded from the steel backing of the bearing. Temperatures may increase to the point where the steel backing turns blue from overheating.

6   Driving habits can have a definite effect on bearing life. Lugging the engine (labouring at low rpm) puts very high loads on bearings, which tends to squeeze out the oil film. These loads cause the bearings to flex, which produces fine cracks in the bearing face (fatigue failure). Eventually the bearing material will loosen in pieces and tear away from the steel backing. Short trip driving leads to corrosion of bearings because insufficient engine heat is produced to drive off the condensed water and corrosive gases. These products collect in the engine oil, forming acid and sludge. As the oil is carried to the engine bearings, the acid attacks and corrodes the bearing material.

7   Incorrect bearing installation during engine assembly will lead to bearing failure as well. Tight fitting bearings leave insufficient bearing oil clearance and will result in oil starvation. Dirt or foreign particles trapped behind a bearing insert result in high spots on the bearing which lead to failure.

23   Engine overhaul – reassembly sequence

1   Before beginning engine reassembly, make sure you have all the necessary new parts, gaskets and seals as well as the following items on hand:

Common hand tools
A 1/2-inch drive torque wrench
Piston ring installation tool
A set of metric Allen wrenches
Piston ring compressor
Short lengths of hose to fit over connecting rod bolts
Plastigage
Feeler gauges
A fine-tooth file
New engine oil
Engine assembly lube or moly-base grease
RTV-type gasket sealant
Anaerobic-type gasket sealant
Thread locking compound

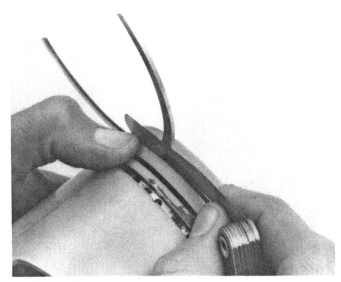

20.10 Check the ring side clearance with a feeler gauge at several points around the groove

20.11 Measure the piston diameter at a 90° angle to the piston pin

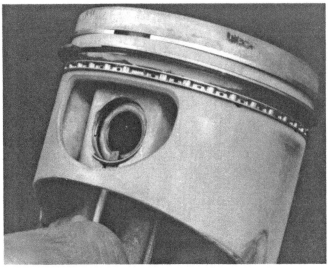

20.12a Pry the snap-rings out of the grooves with a screwdriver

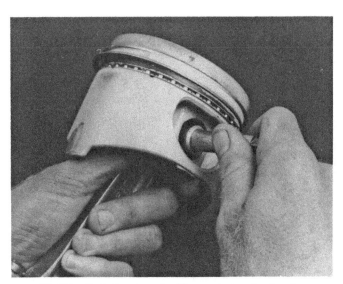

20.12b Push the pin out with a socket and extension

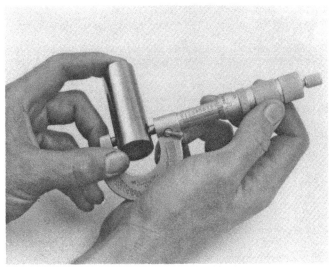

20.12c Measure the pin for wear

21.2 Measure the diameter of each crankshaft journal at several points to detect taper and out-of-round conditions

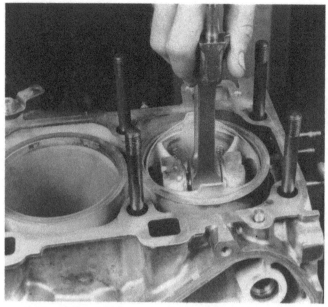

24.3   When checking piston ring end gap, the ring must
be square in the cylinder bore. This is done by pushing the
ring down with the top of a piston

24.4   With the ring square in the cylinder, measure the
end gap with a feeler gauge

24.5   If the end gap is too small, clamp a file in a vise and file
the ring ends (from the outside in only) to enlarge the gap slightly

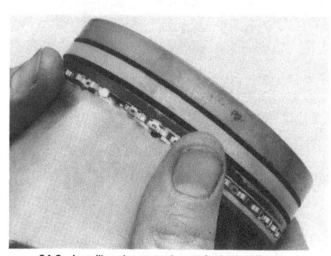

24.9   Installing the spacer/expander in the oil control
ring groove

2   In order to save time and avoid problems, engine reassembly must
be done in the following general order:

> Piston rings
> Crankshaft and main bearings
> Piston/connecting rod assemblies
> Oil pickup
> Balance shafts
> Oil cooler
> Camshaft assembly
> Cylinder head
> Oil pump
> Oil pan
> Camshaft and balance shaft sprockets and belts
> Front cover
> Flywheel/driveplate
> Intake and exhaust manifolds

24   Piston rings — installation

Refer to illustrations 24.3, 24.4, 24.5, 24.9, 24.12 and 24.13

1   Before installing the new piston rings, the ring end gaps must be
checked. It is assumed that the piston ring side clearance has been
checked and verified correct (Section 17).
2   Lay out the piston/connecting rod assemblies and the new ring
sets so the ring sets will be matched with the same piston and cylinder
during the end gap measurement and engine assembly.
3   Insert the top (number one) ring into the first cylinder and square
it up with the cylinder walls by pushing it in with the top of the piston
(see illustration). The ring should be near the bottom of the cylinder,
at the lower limit of ring travel.
4   To measure the end gap, slip feeler gauges between the ends of
the ring until a gauge equal to the gap width is found **(see illustration)**.
The feeler gauge should slide between the ring ends with a slight
amount of drag. Compare the measurement to the Specifications. If
the gap is larger or smaller than specified, double-check to make sure
that you have the correct rings before proceeding.
5   If the gap is too small, it must be enlarged or the ring ends may
come in contact with each other during engine operation, which can
cause serious damage to the engine. The end gap can be increased
by filing the ring ends very carefully with a fine file. Mount the file in
a vise equipped with soft jaws, slip the ring over the file with the ends
contacting the file face and slowly move the ring to remove material
from the ends. When performing this operation, file only from the outside in **(see illustration)**.

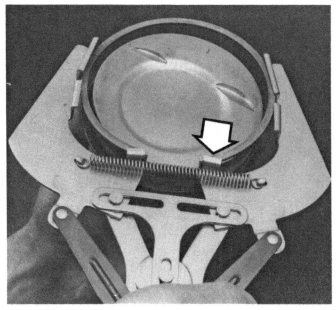

24.12 Installing the compression rings with a ring expander — the number on the top ring (arrow) must face up

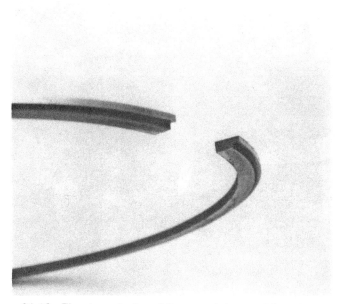

24.13 The stepped edge of the second ring must face down

6 Repeat the procedure for each ring that will be installed in the first cylinder and for each ring in the remaining cylinders. Remember to keep rings, pistons and cylinders matched up.

7 Once the ring end gaps have been checked/corrected, the rings can be installed on the pistons.

8 The oil control ring (lowest one on the piston) is installed first. It is composed of three separate components.

9 Slip the spacer/expander into the groove (see illustration). If an anti-rotation tang is used, make sure it is inserted into the drilled hole in the ring groove. Next, install the lower side rail. Do not use a piston ring installation tool on the oil ring side rails, as they may be damaged. Instead, place one end of the side rail into the groove between the spacer/expander and the ring land, hold it firmly in place and slide a finger around the piston while pushing the rail into the groove. Next, install the upper side rail in the same manner.

10 After the three oil ring components have been installed, check to make sure that both the upper and lower side rails can be turned smoothly in the ring groove.

11 The number two (middle) ring is installed next. It is stamped with a number which must face up. **Note:** *Always follow the instructions printed on the ring package or box — different manufacturers may require different approaches. Do not mix up the top and middle rings, as they have different cross sections.*

12 Use a piston ring installation tool and make sure that the identification mark is facing the top of the piston, then slip the ring into the middle groove on the piston (see illustration). Do not expand the ring any more than is necessary to slide it over the piston.

13 Install the number one (top) ring in the same manner. Make sure the chamfered groove is facing down (see illustration). Be careful not to confuse the number one and number two rings.

14 Repeat the procedure for the remaining pistons and rings.

25 Rear main oil seal — installation

Refer to illustrations 25.2 and 25.3

1 Lubricate the inner circumference of the new seal with clean engine oil or assembly lube.

2 Place the seal in place in the cavity and tap it evenly into the bore until the outer edge is flush with the block surface using a suitable large diameter pipe, socket or similar tool and a block of wood (see illustration).

3 After the seal is in place tap around the edges with a plastic hammer to make sure it is flush (see illustration).

25.2 The rear main oil seal must be tapped into the bore with a large diameter tool. A ring compressor and a block of wood was used here

25.4 Tap around the seal with a plastic hammer to make sure it is seated

26.3   Wipe the bearing surfaces clean before installing the bearings

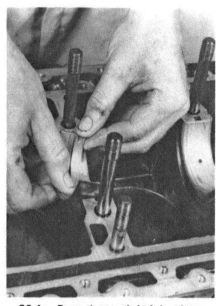

26.4a   Press the crankshaft bearings into the saddles in the block

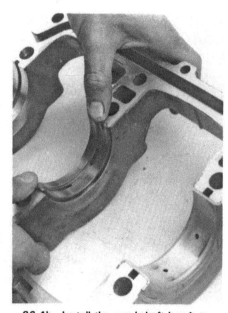

26.4b   Install the crankshaft bearings into the lower crankcase section

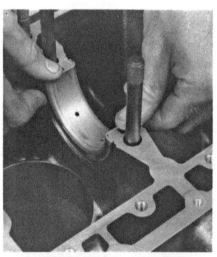

26.5   The flanged thrust bearing is installed in the center saddle

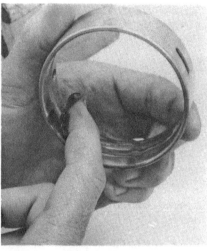

26.7a   Lubricate the inner surface of the front bearing . . .

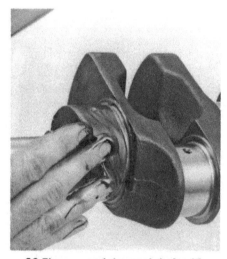

26.7b   . . . and the crankshaft with assembly lube

26.7c   Slide the bearing on with the dowel hole (arrow) facing forward

26.7d   The hole in the front bearing must fit over the dowel in the block

26.9   Lay the Plastigage strip (arrow) on the main bearing journal, parallel to the crankshaft centerline

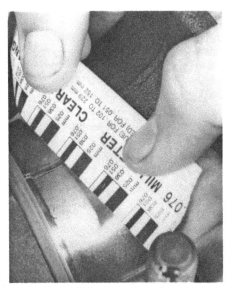

26.13 Compare the width of the crushed Plastigage to the scale on the container to determine the main bearing oil clearance

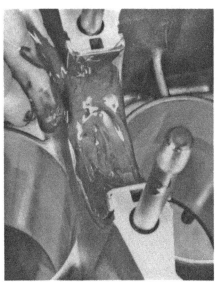

26.16 Be sure to lubricate the thrust faces as well as the bearing surface of the thrust bearing

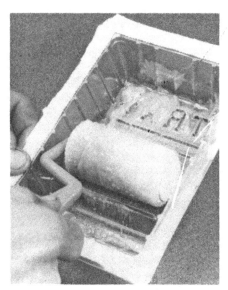

26.18a Apply the sealant to the roller . . .

26 Crankshaft — installation and main bearing oil clearance check

Refer to illustrations 26.3, 26.4a, 26.4b, 26.5, 26.7a, 26.7b, 26.7c, 26.7d, 26.9, 26.13, 26.16, 26.18a, 26.18b and 26.21

1 Crankshaft installation is the first step in engine reassembly. It is assumed at this point that the engine block and crankshaft have been cleaned, inspected and repaired or reconditioned. These engines are unique in that the lower crankcase section incorporates the main bearing caps.

2 Position the engine with the bottom facing up.

3 If they are still in place, remove the old bearing inserts from the block and the lower crankcase section. Wipe the main bearing surfaces of the block and caps with a clean, lint free cloth **(see illustration)**. They must be kept spotlessly clean.

4 Clean the back sides of the new main bearing inserts and lay one bearing half in each main bearing saddle in the block **(see illustration)**. Lay the other bearing half from each bearing set in the lower crankcase main bearing cap **(see illustration)**. Make sure the tab on the bearing insert fits into the recess in the block or cap. Also, the oil holes in the block must line up with the oil holes in the bearing insert. Do not hammer the bearing into place and do not nick or gouge the bearing faces. No lubrication should be used at this time.

5 The flanged thrust bearing must be installed in the center cap and saddle **(see illustration)**.

6 Clean the faces of the bearings in the block and the crankshaft main bearing journals with a clean, lint free cloth.

7 Once you are certain that the crankshaft is clean, lubricate the contact surface of the front bearing and the corresponding surface on the front of the crankshaft with assembly lube **(see illustrations)**. Install the bearing on the crankshaft with the dowel hole facing forward **(see illustration)**. Carefully lay the crankshaft in position (an assistant would be very helpful here) in the main bearings. Make sure the hole in the front bearing mates with the dowel in the block **(see illustration)**.

8 Before the crankshaft can be permanently installed, the main bearing oil clearance must be checked.

9 Trim several pieces of the appropriate size of Plastigage (they must be slightly shorter than the width of the main bearings) and place one piece on each crankshaft main bearing journal, parallel with the journal axis **(see illustration)**.

10 Clean the faces of the bearings in the lower crankcase section and install the lower section. Do not disturb the Plastigage.

11 Install the bolts and tighten them in the specified sequence, in three steps, to the specified torque. Do not rotate the crankshaft at any time during this operation.

12 Remove the bolts and carefully lift off the lower section. Do not

26.18b . . . and roll it onto the lower crankcase section surface

disturb the Plastigage or rotate the crankshaft.

13 Compare the width of the crushed Plastigage on each journal to the scale printed on the Plastigage container to obtain the main bearing oil clearance **(see illustration)**. Check the Specifications to make sure it is correct.

14 If the clearance is not as specified, the bearing inserts may be the wrong size (which means different ones will be required). Before deciding that different inserts are needed, make sure that no dirt or oil was between the bearing inserts and the caps or block when the clearance was measured. If the Plastigage was wider at one end than the other, the journal may be tapered (refer to Section 21).

15 Carefully scrape all traces of the Plastigage material off the main bearing journals and/or the bearing faces. Do not nick or scratch the bearing faces.

16 Carefully lift the crankshaft out of the engine. Clean the bearing faces in the block, then apply a thin, uniform layer of clean moly-base grease or engine assembly lube to each of the bearing surfaces. Be sure to coat the thrust faces as well as the journal face of the thrust bearing **(see illustration)**.

17 Make sure the crankshaft journals are clean, then lay the crankshaft back in place in the block. Clean the faces of the bearings in the lower section, then apply lubricant to them.

18 Apply Loctite 574 anaerobic flange sealant (available at your dealer) to the contact surfaces of the oil gallery groove and the oil seal on the lower section using a close nap roller **(see illustrations)**.

26.21   Lower section/crankshaft bearing nut tightening sequence

19 Place the lower section on the cylinder block immediately because the sealant will dry within ten minutes. Install the nuts.
20 Install the oil pickup/drain tube assembly.
21 Tighten the lower cylinder block section nuts in the specified sequence **(see illustration)**.
22 Install the remaining lower section nuts and bolts and tighten them to the specified torque.
23 On manual transmission equipped models, install a new pilot bearing in the end of the crankshaft (see Chapter 8).
24 Rotate the crankshaft a number of times by hand to check for any obvious binding.
25 The final step is to check the crankshaft end play with a feeler gauge or a dial indicator as described in Section 16. The end play should be correct if the crankshaft thrust faces are not worn or damaged and new bearings have been installed.
26 Refer to Section 25 and install the new rear main oil seal.

---

27 Piston/connecting rod assembly — installation and rod bearing oil clearance check

---

Refer to illustrations 27.5, 27.6, 27.9 and 27.12

1 Before installing the piston/connecting rod assemblies the cylinder walls must be perfectly clean and the crankshaft must be in place.
2 Remove the connecting rod cap from the end of the number one connecting rod. Remove the old bearing inserts and wipe the bearing surfaces of the connecting rod and cap with a clean, lint free cloth. They must be kept spotlessly clean.
3 Clean the back side of the new upper bearing half, then lay it in place in the connecting rod. Make sure that the tab on the bearing fits into the recess in the rod. Do not hammer the bearing insert into place and be very careful not to nick or gouge the bearing face. Do not lubricate the bearing at this time.
4 Clean the back side of the other bearing insert and install it in the

rod cap. Again, make sure the tab on the bearing fits into the recess in the cap, and do not apply any lubricant. It is critically important that the mating surfaces of the bearing and connecting rod are perfectly clean and oil free when they are assembled.
5 Position the piston ring gaps at 120° intervals around the piston, then slip a section of plastic or rubber hose over each connecting rod cap bolt **(see illustration)**.
6 Lubricate the piston and rings with clean engine oil **(see illustration)** and attach a piston ring compressor to the piston. Leave the skirt protruding about 1/4-inch to guide the piston into the cylinder. The rings must be compressed until they are flush with the piston.
7 Rotate the crankshaft until the number one connecting rod journal is at BDC (bottom dead center) and apply a coat of engine oil to the cylinder walls.
8 With the arrow on top of the piston facing to the front of the engine, gently insert the piston/connecting rod assembly into the number one cylinder bore and rest the bottom edge of the ring compressor on the engine block. Tap the top edge of the ring compressor to make sure it is contacting the block around its entire circumference.
9 Carefully tap on the top of the piston with the end of a wooden hammer handle **(see illustration)** while guiding the end of the connecting rod into place on the crankshaft journal. The piston rings may try to pop out of the ring compressor just before entering the cylinder bore, so keep some downward pressure on the ring compressor. Work slowly, and if any resistance is felt as the piston enters the cylinder, stop immediately. Find out what is hanging up and fix it before proceeding. Do not, for any reason, force the piston into the cylinder, as you will break a ring and/or the piston.
10 Once the piston/connecting rod assembly is installed, the connecting rod bearing oil clearance must be checked before the rod cap is permanently bolted in place.
11 Cut a piece of the appropriate size Plastigage slightly shorter than the width of the connecting rod bearing and lay it in place on the number one connecting rod journal, parallel with the journal axis. It must not cross the oil hole in the journal.

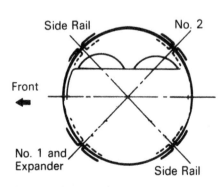

27.5   Piston ring end gap positioning diagram

27.6   Dip the piston into oil to lubricate the rings

27.9   The piston can be driven (gently) into the cylinder bore with of a wooden hammer handle

27.12 The code numbers on the connecting rod and cap must be on the same side

28.1a Tap the seal fully into the oil pump bore using a seal driver . . .

28.1b . . . or a block of wood

28.2 Squirt oil on to the full circumference of the seal

28.3 Check the bolts for length while holding the pump in place

28.4 Apply sealant along the contact surface and around the bolt holes

12 Clean the connecting rod cap bearing face, remove the protective hoses from the connecting rod bolts and install the rod cap. Make sure the code number on the cap is on the same side as the number on the connecting rod (see illustration). Install the nuts and tighten them to the specified torque. Note: *Two different rod cap nuts are used, one with a flat face where the nut contacts the rod cap, and one with a ribbed face. These nuts have different torque values, so be sure to check which nuts your engine has before tightening the rod cap nuts.* Do not rotate the crankshaft at any time during this operation.

13 Remove the rod cap, being very careful not to disturb the Plastigage. Compare the width of the crushed Plastigage to the scale printed on the Plastigage container to obtain the oil clearance. Compare it to the Specifications to make sure the clearance is correct. If the clearance is not as specified, the bearing inserts may be the wrong size (which means different ones will be required). Before deciding that different inserts are needed, make sure that no dirt or oil was between the bearing inserts and the connecting rod or cap when the clearance was measured. Also, recheck the journal diameter. If the Plastigage was wider at one end than the other, the journal may be tapered (refer to Section 21).

14 Carefully scrape all traces of the Plastigage material off the rod journal and/or bearing face. Be very careful not to scratch the bearing — use your fingernail or a piece of hardwood. Make sure the bearing faces are perfectly clean, then apply a uniform layer of clean moly-base grease or engine assembly lube to both of them. You will have to push the piston into the cylinder to expose the face of the bearing insert in the connecting rod — be sure to slip the protective hoses over the rod bolts first.

15 Slide the connecting rod back into place on the journal, remove the protective hoses from the rod cap bolts, install the rod cap and tighten the nuts to the specified torque.

16 Repeat the entire procedure for the remaining piston/connecting rod assemblies. Keep the back sides of the bearing inserts and the inside of the connecting rod and cap perfectly clean when assembling them. Make sure you have the correct piston for the cylinder and that the arrow on the piston faces to the front of the engine when the piston is installed. Remember, use plenty of oil to lubricate the piston before installing the ring compressor.

17 After all the piston/connecting rod assemblies have been properly installed, rotate the crankshaft a number of times by hand to check for any obvious binding.

28 Oil pump — assembly and installation

Refer to illustrations 28.1a, 28.1b, 28.2, 28.3, 28.4 and 28.6

1 Place the new seal into the pump and tap it evenly into the bore using a hammer and a seal driver or block of wood (see illustrations).
2 Lubricate the contact surfaces of the seal with clean engine oil (see illustration).
3 The oil pump bolts are of varying lengths, so before installation, place the pump on the studs and check the bolts for length to determine where they will be installed (see illustration).
4 Apply a bead of Loctite 574 anaroebic sealant around the pump contact surfaces (see illustration).

28.6   Insert the sleeve into the pump

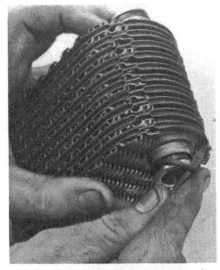

29.1a   Use a screwdriver to remove the O-ring

29.1b   Press the new O-ring into the groove

29.1c   Pry the old seals off with a screwdriver

29.2   Press the mount securely into the block recess

29.3   Place the cooler on the block and press down to seat the mount

5   Place the pump in position on the block and install the bolts.
6   Push the sleeve into position and seat it by moving it back and forth **(see illustration)**.
7   Tighten the bolts to the specified torque.
8   Lubricate the O-ring with clean engine oil and slide it onto the oil pump, followed by the thrust washer.
9   Install the key in the pump shaft.

---

29   Oil cooler — installation

Refer to illustrations 29.1a, 29.1b, 29.1c, 29.2, 29.3, 29.4, 29.5, 29.9, 29.13a, 29.13b, 29.13c, and 29.14

Note: *1987 and later models have a modified oil cooler housing; the new style housing does not have a raised boss at the bottom of the housing. When overhauling a 1986 or earlier model engine it is recommended that the new style housing, and the 1987 and later installation procedure, be used.*

1   Install new O-rings and seals on the cooler element **(see illustrations)**.
2   Install a new rubber mount in the recess in the block **(see illustration)**.

1986 and earlier models

3   Insert the cooler into position and press it into place to seat the rubber mount in the block **(see illustration)**.
4   Install the new seal and O-ring in the housing **(see illustration)**.
5   Coat the groove in the housing with RTV sealant and then press the new rubber gasket securely into the groove **(see illustration)**.
6   Lower the housing in place over the cooler.
7   Install the bolts and tighten them securely.

1987 and later models

8   Install the cooler and plastic washer into the housing, pushing down to make sure it is fully seated.
9   Lay a straightedge across the housing (without the housing gasket in place) and measure the distance from the housing mating surface to the contact surface of the guide pin **(see illustration)**. This clearance must be shimmed to a distance of 0.00 +/- 0.010-inch (0.25 mm). Aluminum shims made specifically for this purpose are available at your dealer parts department and some auto parts stores specializing in foreign cars.
10 Once the clearance has been properly set, install the cooler and housing onto the engine block, using a new gasket. Tighten the bolts securely.

29.4 Press the seal and the O-ring securely into their grooves

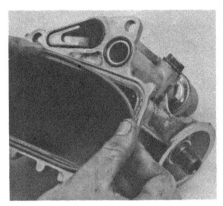

29.5 Hold the gasket with one hand while pressing it into the groove with the thumb of the other hand

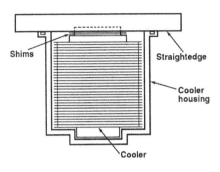

29.9 Cutaway of oil cooler housing showing relationship of shims to mating surface of housing

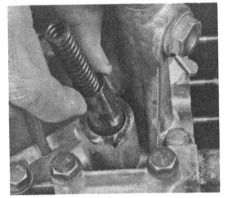

29.13a Lower the valve and spring into the housing

29.13b Compress the spring with the plug . . .

29.13c . . . and screw the plug in

All models

11 Lubricate the relief valve and spring with clean engine oil or assembly lube.

12 Install a new gasket on the relief valve plug.

13 Insert the valve and spring, install the plug and tighten it securely **(see illustrations).**

14 Place a new gasket on the oil filter boss **(see illustration).**

29.14 Slide the new gasket over the oil filter boss

---

30 Initial start-up and break-in after overhaul

---

1 Once the engine has been installed in the vehicle, double-check the engine oil and coolant levels.

2 With the spark plugs out of the engine and the ignition switch feed wires (the two smaller wires coming from the ignition coil) disconnected, crank the engine until oil pressure registers on the gauge.

3 Install the spark plugs, hook up the plug wires and reconnect the ignition coil wires.

4 Start the engine. It may take a few moments for the gasoline to reach the injectors, but the engine should start without a great deal of effort.

5 After the engine starts, it should be allowed to run at approximately 2000 rpm until it warms up to normal operating temperature. While the engine is warming up, make a thorough check for oil and coolant leaks.

6 Shut the engine off and recheck the engine oil and coolant levels.

7 Drive the vehicle to an area with minimum traffic, accelerate at full throttle from 30 to 50 mph, then allow the vehicle to slow to 30 mph with the throttle closed. Repeat the procedure 10 or 12 times. This will load the piston rings and cause them to seat properly against the cylinder walls. Check again for oil and coolant leaks.

8 Drive the vehicle gently for the first 500 miles (no sustained high speeds) and keep a constant check on the oil level. It is not unusual for an engine to use oil during the break-in period.

9 At approximately 500 to 600 miles, change the oil and filter.

10 For the next few hundred miles, drive the vehicle normally. Do not pamper it or abuse it.

11 After 2000 miles, change the oil and filter again and consider the engine fully broken in.

Chapter 3
Cooling, heating and air conditioning systems

Contents

Specifications

Torque specifications

	Ft-lbs	Nm
Balance shaft idler bolt	33	45
Coolant temperature switch	22	30
Camshaft idler bolt	32	45
Crankshaft sprocket bolt	151	210
Water pump bolts	6	8
Drivebelt cover bolts	6	8
Camshaft sprocket cover bolt	6	8
Balance shaft sprocket bolt	32	45
Radiator drain plug	4	5
Engine block coolant drain plug	14	20
Tensioning roller nut	32	45

1 General information

The cooling system consists of a radiator, oil cooler, coolant recovery reservoir, thermostatically controlled cooling fan, thermostat and housing and engine-driven water pump.

The radiator is aluminum with plastic tanks. Radiators on automatic transmission equipped vehicles are equipped with additional inlet and outlet fittings which circulate transmission fluid through an oil cooler inside the right hand tank.

The radiator does not have a cap. The plastic coolant recovery reservoir located in the engine compartment immediately behind the radiator is connected to the radiator by hoses. A pressure cap on the reservoir allows a buildup of pressure in the cooling system. This pressure allows the boiling point of the coolant to be raised.

The electric cooling fan is located in a shroud assembly mounted on the radiator. Air conditioning equipped models have two fans. The cooling fan is activated by a temperature switch threaded into the radiator. **Warning:** *The electric cooling fan on these models can activate at any time, even when the ignition is in the Off position. Disconnect the fan motor or negative battery cable when working in the vicinity of the fan.*

The water pump is mounted on the front of the engine block under

a cover and is driven off the camshaft drivebelt. The pump draws coolant from the radiator and circulates it through water jackets in the engine block, oil cooler and cylinder head.

The oil cooler is located on the right side of the engine and incorporates the oil filter housing. Engine oil circulating through the cooler housing is brought up to the coolant temperature. This aids in engine warm up and maintains the oil supply at a constant temperature.

Engine coolant travels between the radiator and the engine through hoses. Some of the hot coolant is diverted through the heater core. A fan blows the heated air through several heating ducts located in the dash.

Air conditioning is available as an option. The compressor is located on the left front of the engine and is turned by a crankshaft pulley driven polyrib drivebelt tensioned by a pressure rod. The refrigerant reservoir tank is located in the left front corner of the engine compartment.

2 Antifreeze — general information

Warning: *Do not allow antifreeze to come in contact with your skin or painted surfaces of the vehicle. Flush contacted areas immediately* *with plenty of water. Do not store new coolant or leave old coolant lying around where it is easily accessible to children and pets, because they are attracted by its sweet taste. Ingestion of even a small amount can be fatal. Wipe up garage floor and drip pan coolant spills immediately. Keep antifreeze containers covered and repair leaks in your cooling system as soon as possible.*

The cooling system should be filled with a water/ethylene glycol based antifreeze solution which will prevent freezing down to at least −20°F. It also provides protection against corrosion and increases the coolant boiling point.

The cooling system should be drained, flushed and refilled at least every other year (see Chapter 1). The use of antifreeze solutions for periods of longer than two years is likely to cause damage and encourage the formation of rust and scale in the system.

Before adding antifreeze to the system, check all hose connections and retorque the cylinder head nuts, because antifreeze tends to search out and leak through very minute openings.

The exact mixture of antifreeze to water which you should use depends on the relative weather conditions. The mixture should contain at least 50 percent antifreeze, but should never contain more than 70 percent antifreeze.

3 Thermostat — removal and installation

Refer to illustrations 3.3, 3.4, 3.5a and 3.5b

Warning: *The electric cooling fan on these models can activate at any time, even when the ignition is in the Off position. Disconnect the fan motor or negative battery cable when working in the vicinity of the fan. The engine must be completely cool before beginning this procedure.*

1 The thermostat housing is located on the outlet of the water pump at the lower radiator hose-to-pump junction. It is held in place by a snap-ring so it will be necessary to obtain snap-ring pliers before beginning work.
2 Drain the cooling system (Chapter 1).
3 Working from below the engine compartment, remove the lower radiator hose **(see illustration)**.
4 Remove the snap-ring **(see illustration)**.
5 Withdraw the thermostat from the water pump housing **(see illustration)**. Be prepared for a rush of coolant as the thermostat is removed. If it did not come out with the thermostat, make sure the thermostat gasket is removed from the water pump **(see illustration)**.

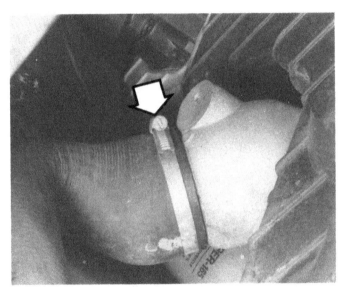

3.3   The thermostat is located in the water pump outlet and is accessible after removing the clamp (arrow) and radiator hose

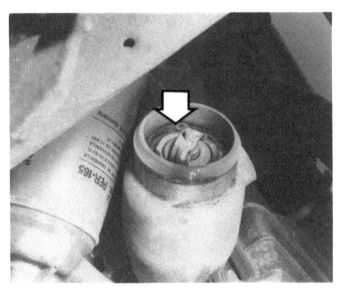

3.4   The thermostat can be removed after the snap-ring (arrow) is removed

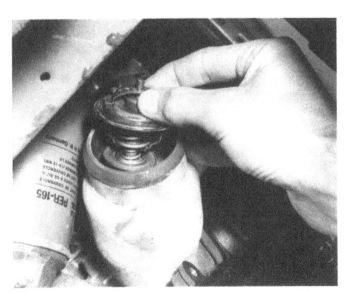

3.5a   Note how the thermostat is installed (which end is facing into the engine), then pull it out of the water pump housing

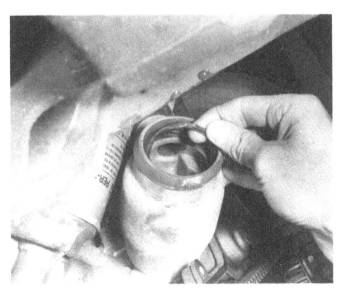

3.5b   The gasket may come off the thermostat — be sure to retrieve it from the pump housing

6   Check the gasket for damage and distortion. Replace it with a new one if necessary. Install the gasket on the thermostat.
7   Insert the thermostat into position, making sure it seats securely.
8   Install the snap-ring in the groove.
9   The remainder of installation is the reverse of removal.

4   Thermostat — check

1   If you suspect a malfunctioning thermostat, it should be replaced with a new one. However, since the thermostat for this particular vehicle is relatively expensive it can be checked first to verify that it is defective.
2   Remove the thermostat (Section 3).
3   Inspect the thermostat for excessive corrosion and damage. Replace it with a new one if either of these conditions is noted.
4   Place the thermostat in a container filled with water. You will need a cooking thermometer to measure water temperature.
5   Heat the water to the temperature stamped on the bottom of the thermostat (remember that this will usually be in centigrade). When submerged in hot water heated to its rated operating temperature, the valve should open all the way.
6   Allow the water to cool until it's about 10 degrees below the temperature indicated on the thermostat. At this temperature, the thermo-

stat valve should close completely.
7   Reinstall the thermostat if it operates properly. If it does not, purchase a new thermostat of the same temperature rating.

5   Coolant reservoir — removal and installation

Refer to illustrations 5.2, 5.3a, 5.3b and 5.4
Warning: *The electric cooling fan on these models can activate at any time, even when the ignition is in the Off position. Disconnect the fan motor or negative battery cable when working in the vicinity of the fan.*

1   Drain the cooling system to below the level of the reservoir (Chapter 1).
2   Disconnect the accessible tank hoses **(see illustration)**.
3   Loosen the bolt at the base of the reservoir and remove the two bolts retaining it to the fender **(see illustrations)**.
4   Lift the reservoir up to provide access to the remaining hose. Disconnect the hose and remove the tank from the vehicle **(see illustration)**.
5   Inspect the reservoir for cracks and distortion. Check the cap to make sure it is sealing properly. If the reservoir is damaged, replace it.
6   To install it, connect the hose and lower the reservoir into place. Connect the remaining hoses and install the retaining bolts.

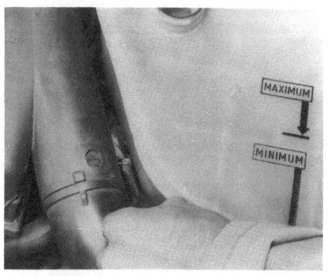

5.2   Use a screwdriver to loosen the reservoir hose clamps

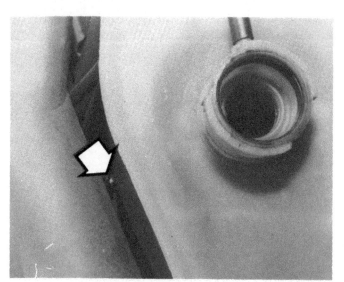

5.3a   The coolant reservoir is held in place by a bolt at the bottom (arrow) . . .

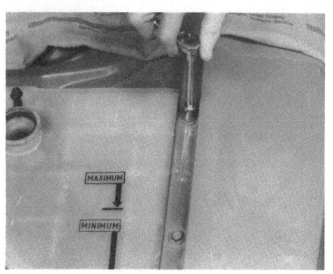

5.3b   . . . and two bolts at the fender mount

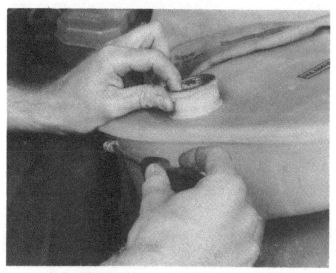

5.4   With the reservoir raised, the lower hose can be disconnected

6 Coolant temperature switch — check

Refer to illustration 6.2

1   Inspect the coolant switch for corrosion and damage. Replace it with a new one if either of these conditions is noted.
2   Disconnect the wires from the switch **(see illustration)**.
3   With the switch cold (do not run the engine for at least 5 or 6 hours prior to this check) check to make sure there is no continuity between the two switch terminals.
4   Start the engine and let it run until normal operating temperature is reached. As the engine continues to idle, the coolant temperature will increase until it is greater than the temperature stamped on the switch. At that point, the switch should have continuity. DO NOT allow the engine to get too hot — remember, the fan will not operate with the coolant switch disconnected.
5   If the switch fails either test, replace it with a new one.

7 Coolant temperature switch — removal and installation

1   Disconnect the negative cable at the battery. Place the cable out of the way so it cannot accidentally come in contact with the negative terminal of the battery, as this would once again allow power into the electrical system of the vehicle.
2   Remove the fan assembly (Section 9).
3   The coolant switch is located on the left side of the radiator, just below the upper radiator hose **(see illustration 6.2)**. Drain the radiator (Chapter 1).
4   Unplug the wires and unscrew the switch with a wrench or socket.
5   To install, screw the coolant switch into the radiator and tighten it to the specified torque.
6   The remainder of installation is the reverse of removal.

8 Electric cooling fan circuit — check

Refer to illustrations 8.4 and 8.5

Circuit operation

1   The cooling fan motor(s) are operated by the fan relay. When the coolant exceeds normal operating temperature, the coolant temperature switch closes, turning the fan on.

System test

2   Run the engine at a fast idle until the fan comes on.
3   If the fan does not come on when the engine is hot, check the fan motor fuse (Chapter 12).
4   To check the fan circuit, fuse and relay for proper operation, bypass the temperature switch. Unplug the switch wires and connect a jumper wire between them **(see illustration)**. The motor(s) should run with the temperature switch bypassed, indicating that the switch is faulty.
5   To check the motor(s), unplug the connector, connect a jumper wire between one fan motor terminal and a good ground and apply battery power to the other terminal by connecting a jumper wire to the positive battery post **(see illustration)**.
6   If the fan runs all the time, even when the engine is cold, check the coolant temperature switch, replacing it if necessary (Sections 6 and 7).

6.2   After unplugging the wires, the coolant temperature switch (arrow) can be checked with an ohmmeter

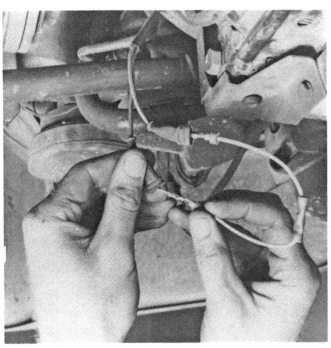

8.4   The fan motor(s) should run when the temperature switch is bypassed by connecting a jumper wire between the two switch leads

8.5   Connect one fan motor terminal to a good ground and the other to the battery positive post to check the motor

9   Electric cooling fan — removal and installation

Refer to illustrations 9.2, 9.3, 9.4, 9.5a, 9.5b, 9.5c and 9.6

1   Disconnect the negative cable at the battery. Place the cable out of the way so it cannot accidentally come in contact with the negative terminal of the battery, as this would once again allow power into the electrical system of the vehicle.

2   Remove the bracket between the air cleaner housing and the upper radiator support **(see illustration)**.

3   In the engine compartment, cut the wiring harness straps attached to the fan shroud **(see illustration)**.

4   Remove the three bolts retaining the fan shroud assembly to the top of the upper radiator support **(see illustration)**.

5   From under the engine compartment, unplug the fan motor connector and disconnect the harness from the shroud, cutting any plastic retaining straps **(see illustrations)**.

6   Remove the three bolts and lower the fan assembly from the vehicle **(see illustration)**.

7   Installation is the reverse of removal. Be sure to replace the plastic wiring harness straps with new ones.

10   Radiator — removal and installation

Refer to illustrations 10.4, 10.6a, 10.6b and 10.8
Warning: *The engine must be completely cool before beginning this procedure.*

1   Disconnect the negative cable at the battery. Place the cable out of the way so it cannot accidentally come in contact with the negative terminal of the battery, as this would once again allow power into the electrical system of the vehicle.

2   Drain the cooling system (Chapter 1).

3   Remove the fan assembly (Section 9).

4   Remove the nuts and lift off the two clamps retaining the top of the radiator **(see illustration)**.

5   Unplug the coolant switch wires.

6   Disconnect the radiator hoses **(see illustrations)**.

9.2   The air cleaner bracket is held in place by two bolts and two nuts (arrows)

9.3   Use wire cutters to cut the harness straps attached to the fan shroud

9.4   Use a socket with an extension to remove the three upper fan shroud bolts

9.5a   Unplug the fan motor connector, . . .

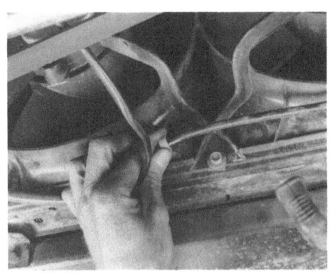

9.5b   . . . and disconnect the harness . . .

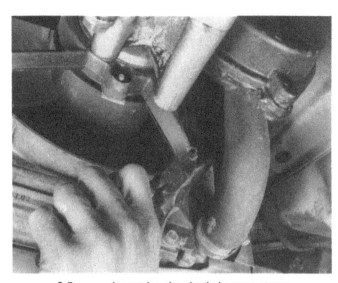

9.5c   . . . by cutting the plastic harness straps

9.6   The three lower fan motor shroud mounting bolts
(arrows) are accessible from under the vehicle

10.4   The upper radiator clamps are held in place
by two nuts (arrows)

10.6a   Loosen the clamp (arrow) and detach the upper
radiator hose — if the hose is stuck, grasp it with
a large pair of Channelock pliers, twist it on the
fitting, then pull it off

10.6b   The right side has two hoses attached
to it (arrows)

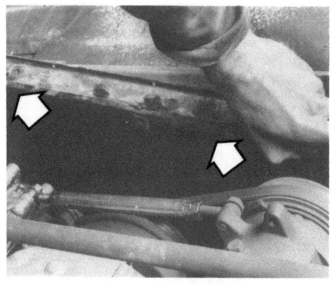

10.8 Lift up on the radiator to disengage it from the mount, then swing the bottom to the rear and lower it from the vehicle

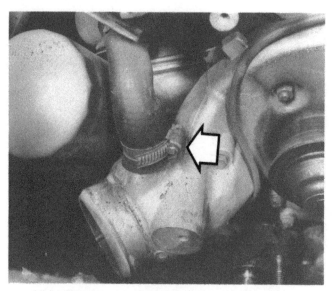

12.6   The water pump has two hoses connected to it; the lower radiator hose (already removed here) and the heater hose (arrow)

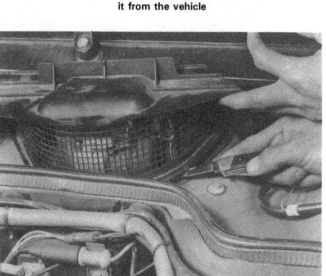

13.2a   Pull up on the motor cover while cutting the adhesive

13.2b   Lift the cover off the motor after the adhesive has been cut

7   On automatic transmission equipped models, disconnect and plug the transmission fluid hoses.

8   Push up on the radiator to disengage it from the rubber mounts, pull the bottom of the radiator to the rear and then lower it from the vehicle **(see illustration)**.

9   Inspect the rubber radiator mounts prior to installation. Replace the mounts if they are cracked or deteriorated.

10   Installation is the reverse of removal. Now is a good time to consider replacing the hoses with new ones (Chapter 1).

11   Water pump — check

1   Water pump failure can cause overheating and serious engine damage because a defective pump will not circulate coolant through the engine.

2   There are three ways to check the operation of the water pump while it is still installed on the engine. If any of the following checks indicate that the pump is defective, replace it with either a new or rebuilt unit.

3   Squeeze the upper radiator hose while the engine is running at nor-

mal operating temperature. If the water pump is working properly, a pressure surge will be felt as the hose is released.

4   If the water pump seal fails, small amounts of coolant will leak out, signalling trouble.

5   If the water pump shaft bearing fails, it will usually emit a squealing sound while it is running. Do not confuse drivebelt slippage, which also makes a squealing sound, with water pump bearing failure.

12   Water pump — removal and installation

Refer to illustration 12.6

Warning: *The engine must be completely cool before beginning this procedure.*

Caution: *This procedure, while straightforward, involves the removal and installation of the camshaft and balance shaft belts. Severe engine damage can result if the camshaft and balance shaft belts or sprockets are not returned to their original positions and the belts properly tensioned after reinstallation. The home mechanic should consider very seriously the scope of this procedure before proceeding.*

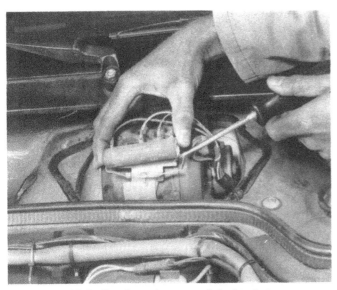

13.3 Pry the resistor out of the bracket
with a screwdriver

13.4a Unplug the motor connector . . .

13.4b . . . and remove the bracket retaining screw

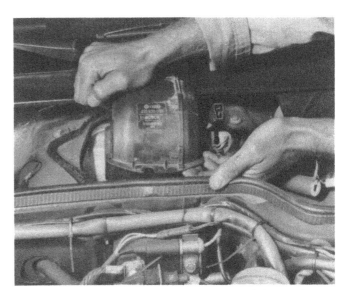

13.5 Lift the blower motor straight up out of the opening

Removal

1 Disconnect the negative cable at the battery. Place the cable out of the way so it cannot accidentally come in contact with the negative terminal of the battery, as this would once again allow power into the electrical system of the vehicle.
2 Drain the cooling system (Chapter 1).
3 Remove the timing belt cover (Chapter 2).
4 Remove the balance shaft and camshaft drivebelts, sprockets and pulleys (Chapter 2).
5 Remove the inner belt cover bolts and detach the cover.
6 In the engine compartment, disconnect the coolant hoses (see illustration) and remove the water pump retaining bolts and nut.
7 Detach the water pump from the engine. It may be necessary to break the gasket seal by removing the thermostat and inserting a suitable bar into the water pump to use as a lever.

Installation

8 Carefully clean the sealing surfaces of the engine and water pump. Prior to installation, coat the gasket with sealant and the water pump bolt threads with locking compound. Place the gasket and water pump in position and install the bolts. Tighten the bolts to the specified torque.
9 Install the inner belt cover.
10 Install the camshaft and balance shaft sprockets and drivebelts (Chapter 2).

11 Install the timing belt cover (Chapter 2).
12 Fill the cooling system with the specified coolant (Chapter 1).

---

13 Heater blower motor — removal and installation

Refer to illustrations 13.2a, 13.2b, 13.3, 13.4a, 13.4b, 13.5 and 13.6

1 Disconnect the negative cable at the battery. Place the cable out of the way so it cannot accidentally come in contact with the negative terminal of the battery, as this would once again allow power into the electrical system of the vehicle.

1983 and 1984 models

2 Pry out the plastic clip at the front edge of the motor cover and then pull up on the cover while carefully running a knife around the base to cut the adhesive (see illustrations).
3 Remove the resistor from the bracket by prying it out with a screwdriver (see illustration).
4 Unplug the motor connector and remove the resistor bracket (see illustrations).
5 Carefully lift the motor up and out of the firewall cavity and remove it from the vehicle (see illustration).

6   To install the motor, lower it into place, engaging the rubber lug on the back (windshield) side with the gasket **(see illustration)**, plug in the connectors, install the bracket and resistor and carefully press the cover and retaining clip into place.

1985 and 1986 models

7   Carefully pull off the fresh air well cover, loosening the adhesive underneath the windshield and then folding the cover up onto the windshield.
8   Remove the six blower mounting screws and unplug the connectors.
9   Raise the blower motor sufficiently to provide access to the vacuum line, disconnect the line and then remove the blower motor.
10   Inspect the sealing cord for damage, replacing it with a new one if necessary.
11   Installation is the reverse of removal.

---

14   Air conditioning system - servicing

Refer to illustration 14.7

Warning: *The air conditioning system is pressurized at the factory and requires special equipment for service and repair. Any work should be left to your dealer or a refrigeration shop. Do not, under any circumstances, disconnect the air conditioning hoses while the system is under pressure.*

1   The following maintenance steps should be performed on a regular basis to ensure that the air conditioner continues to operate at peak efficiency.
   a) Check the tension of the drivebelt and adjust if necessary (Chapter 1).
   b) Inspect the condition of the hoses. Check for cracks, hardening and deterioration. **Warning:** *Do not replace A/C hoses until the system has been discharged by a dealer or air conditioning specialist.*
   c) Inspect the fins of the condenser for leaves, bugs and any other foreign material. A soft brush and compressed air can be used to remove them.
   d) Maintain the correct refrigerant charge.

2   The A/C compressor should be run for about 10 minutes at least once a month. This is particularly important during the winter months because long-term non-use can cause hardening of the internal seals.
3   Because of the complexity of the air conditioning system and the special equipment required to effectively work on it, accurate troubleshooting and repair of the system should be left to a professional mechanic. One probable cause for poor cooling that can be determined by the home mechanic is low refrigerant charge. Should the system lose its cooling ability, the following procedure will help you pinpoint the cause.
4   Warm the engine to normal operating temperature.
5   The hood and doors should be open.
6   Turn the air conditioner On.
7   With the compressor engaged, check the sight glass of the reservoir located in the left front corner of the engine compartment **(see illustration)**.
8   If the glass is clear (occasional bubbles are permissible), the refrigerant level is probably okay. The problem is elsewhere.
9   If foam or a stream of bubbles is visible the system is low on refrigerant. Add refrigerant as follows.
10   Buy an automotive air conditioner recharge kit and hook it up to the low (suction) side in accordance with the kit manufacturer's instructions. Add refrigerant until the reservoir sight glass is clear with occasional bubbles. Allow stabilization time between each addition. **Note:** *Because of recent environmental laws, recharge kits and 12-ounce refrigerant cans may not be available in your area. If this is the case, take the vehicle to a dealer service department or licensed automotive air conditioning technician to have refrigerant added.*

---

15   Air conditioning compressor - removal and installation

Refer to illustrations 15.4, 15.5a, 15.5b and 15.6

Warning: *The air conditioning system must be discharged by a dealer service department or an automotive air conditioning shop before performing this procedure.*

1   Disconnect the negative cable at the battery. Place the cable out of the way so it cannot accidentally come in contact with the negative terminal of the battery, as this would once again allow power into the electrical system of the vehicle.

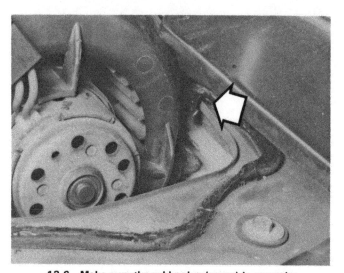

13.6   Make sure the rubber lug (arrow) is securely engaged after the blower motor is lowered into place

14.7   The sight glass (arrow) is located at the top of the reservoir and should be clear if the system is properly charged

2 Raise the vehicle and support it on jackstands.
3 Loosen the tensioner and remove the drivebelt (Chapter 1).
4 Remove the tensioner-to-compressor bolt **(see illustration)**.
5 Remove the compressor mounting bolts **(see illustrations)**.
6 Lower the compressor and disconnect the hose fittings with a wrench **(see illustration)**.
7 Remove the compressor from the vehicle.
8 Installation is the reverse of removal.

16 Air conditioning condenser - removal and installation

Refer to illustrations 16.1a, 16.1b and 16.2

Warning: *The air conditioning system must be discharged by a dealer service department or an automotive air conditioning shop before performing this procedure.*

1 Working through the grille opening, disconnect the air conditioner lines at the condenser using two wrenches to unscrew the fittings **(see illustrations)**.

15.4 Use a wrench to unbolt the belt tensioner from the compressor

15.5a Use a socket and extension to remove the front . . .

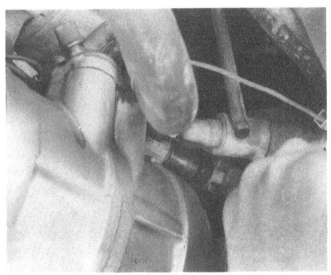

15.5b . . . and rear compressor mounting bolts

15.6 Lower the compressor and unscrew the line connection with a wrench (arrow)

16.1a The condenser line connections (arrows) are accessible through the grille opening

16.1b Disconnect the condenser lines using two wrenches — one to hold the fitting and the other to unscrew the line nut

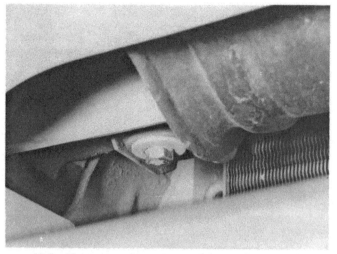

16.2 The two condenser mount bolts can be reached through the grille openings

2   Remove the two mount bolts, also accessible through the grille air intake **(see illustration)**.
3   Carefully move the condenser to the rear and lower it from the vehicle.
4   Installation is the reverse of removal.

---

17   Air conditioning reservoir — removal and installation

Refer to illustration 17.2
Warning: *The air conditioning system must be discharged by a dealer service department or an automotive air conditioning shop before performing this procedure.*

1   Disconnect the negative cable at the battery. Place the cable out of the way so it cannot accidentally come in contact with the negative terminal of the battery, as this would once again allow power into the electrical system of the vehicle.
2   Disconnect the lines at the connections on the reservoir, loosen the bracket screw, then lift the reservoir out of the bracket **(see illustration)**.
3   Installation is the reverse of removal.

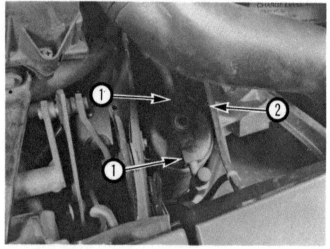

17.2   Disconnect the lines at the reservoir (1), loosen the bracket screw (2) and lift the reservoir out of the bracket

Chapter 4 Fuel and exhaust systems

Contents

Specifications

Voltage and resistance checks

Air flow sensor check

Non-turbocharged engine	
voltage between terminal 9 and ground	8 or more volts
Turbocharged engine	
voltage between terminal 3 and ground	5 volts
Non-turbocharged engine	
voltage between terminal 7 and ground	
sensor door closed	150 to 250 millivolts
sensor door open	8 or more volts
Turbocharged engine	
voltage between terminal 2 and ground	
sensor door closed	250 to 260 millivolts
sensor door open	4.60 volts

Air flow sensor temperature sensor

Resistance between connector terminals 6 and 22	
non-turbocharged engine	1.45 to 3.3 K ohms @ 59 to 86 °F (15 to 30 °C)
turbocharged engine	1.4 to 3.6 K ohms @ 59 to 86 °F (15 to 30 °C)

Engine temperature sensor

Resistance between control unit terminal 13 and ground	
non-turbocharged engine	
59 to 86 °F (15 to 30 °C)	1.45 to 3.3 K ohms
176 °F (80 °C)	280 to 360 ohms
turbocharged engine	
59 to 86 °F (15 to 30 °C)	1.4 to 3.6 K ohms
176 °F (80 °C)	250 to 390 ohms

Throttle switch check (switch removed)

Control unit plug terminal 2 to ground	
throttle closed	0 ohms
throttle open	Infinite ohms
Control unit plug terminal 3 to ground	
throttle closed	Infinite ohms
throttle open	0 ohms
Fuel injector resistance	2 to 3 ohms
Auxiliary air regulator voltage	10 volts
Auxiliary air regulator resistance	20 to 55 ohms

General

Fuel pump output	850 cc/30 seconds
Fuel pressure	
at idle	Approximately 29 psi
vacuum hose disconnected	33 to 39 psi
fuel return hose clamped shut	58 psi

Torque specifications

	Ft-lbs	Nm
Fuel rail cap	16	22

1.1 Digital Engine Electronic (DEE) system component layout

1 Pressure regulator
2 Fuel rail
3 Pressure damper
4 Intake manifold
5 Air flow sensor
6 Throttle switch
7 Engine temperature sensor
8 Fuel injector

2.2 Removing the fuel pump fuse (arrow) will disable the fuel pump so the fuel system can be depressurized

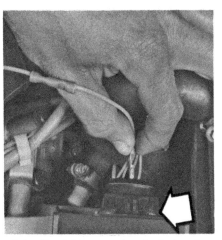

3.3 Check between the terminal and a good ground with a voltmeter. The terminal numbers are visible on the top edge of the connector (arrow)

3.4 Check the throttle position sensor connector to make sure the specified voltage is reaching the sensor

1 General information

Refer to illustration 1.1

Fuel system

The fuel system consists of the fuel tank, an electrically operated fuel pump, an air cleaner assembly and an L-Jetronic fuel injection system controlled by the Digital Engine Electronic (DEE) (also called Digital Motor Electronic) system (Chapter 5).

The DEE system controls the ignition timing and the injection of the fuel/air mixture into the combustion chambers, using information from the oxygen sensor in the exhaust pipe along with air flow and throttle position sensors in the injection system.

Air is drawn into the air cleaner housing and filter, past the air flow sensor, through a pipe to the throttle assembly and throttle position sensor and then through the intake manifold to the cylinders. An auxiliary air valve mounted below the intake manifold supplies additional air when the engine is cold.

Fuel is drawn from the fuel tank by an electric pump mounted adjacent to the tank, through a filter to the fuel rail mounted on the engine. The fuel rail distributes fuel to the individual injectors.

The fuel injectors are mounted between the fuel rail and the intake ports of the engine. Fuel is delivered to the injectors at a constant pressure level, which is maintained by a pressure regulator on the end of the fuel rail. A pressure damper, similar in appearance to the pressure regulator, is mounted on the intake manifold side of the fuel rail. The damper absorbs pulsations in the fuel pressure caused by the operation of the fuel injectors. After passing through the rail the unused fuel is returned to the fuel tank by the pressure regulator.

The fuel injectors are solenoid operated devices controlled by a signal from the DEE control unit. The control unit opens each injector, allowing fuel to spray into the cylinder in accordance with the information supplied to it by the air flow sensor, throttle position sensor, oxygen sensor and temperature sensors. The amount of time the injector is held open determines the fuel/air mixture ratio.

Exhaust system

The exhaust system includes an exhaust manifold, an exhaust pipe fitted with an oxygen sensor, a catalytic converter and a muffler with associated pipes.

The catalytic converter is an emission control device added to the exhaust system to reduce pollutants, notably oxides of nitrogen (NOX) as well as hydrocarbons (HC) and carbon monoxide (CO).

2 Fuel system depressurization

Refer to illustration 2.2

Warning: *Gasoline is extremely flammable, so extra precautions must be taken when working on any part of the fuel system. Do not smoke or allow open flames or bare light bulbs near the work area. Also, do not work in a garage if a natural gas-type appliance with a pilot light is present.*

1 Depressurize the fuel system before working on any fuel system components. Otherwise the pressure which is maintained in the system even when the engine is off could cause fuel to spray when a part of the system is disconnected.

2 Remove the fuel pump fuse (second from right) from the auxiliary fuse box located under the left side of the instrument panel **(see illustration)**.

3 Start the engine and run it until it stops from lack of fuel. Although the pressure will now be released through the injectors into the cylinders, remember when disconnecting any lines that some residual fuel will remain in the system.

3 Fuel injection components — checking and replacement

Refer to illustrations 3.3, 3.4, 3.6, 3.7, 3.11, 3.13, 3.20a, 3.20b, 3.20c, 3.20d, 3.21, 3.22, 3.23, 3.24, 3.25, 3.26, 3.27, 3.28, 3.32, 3.34a, 3.34b, 3.35, 3.36, 3.37, 3.38a, 3.38b, 3.39a, 3.39b, 3.40, 3.41, 3.43 and 3.44

Warning: *Gasoline is extremely flammable, so extra precautions must be taken when working on any part of the fuel system. Do not smoke or allow open flames or bare light bulbs near the work area. Also, do not work in a garage if a natural gas-type appliance with a pilot light is present.*

Caution: *Use only a High Impedance digital voltmeter when performing the check below or serious damage to the electrical components could result.*

Air flow sensor check

1 Remove the air cleaner cover and filter element (Chapter 1).

2 Push the protective boot back on the air flow sensor electrical connector to expose the terminals.

3 Check between the back of terminal 9 on the connector (non-turbocharged models) or terminal 3 (turbocharged models) and a ground with a voltmeter **(see illustration)**. Have an assistant turn the ignition On and compare the voltage reading to the Specifications.

4 If the voltmeter reading is not as specified, unplug the connector

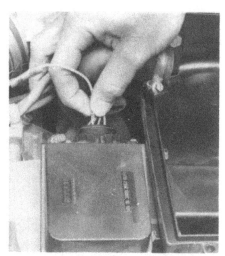

3.6   Hold the air flow sensor door open while checking the voltage at the connector

3.7   Check the air flow sensor temperature sensor voltage at the outer two connector pins (6 and 22)

3.11   Pull the retainer away and then unplug the control unit connector by swinging it out of the unit (the opposite end pivots on the end of the unit)

3.13   Engine temperature sensor location (arrow)

3.20a   After disconnecting the intake air hose, use a screwdriver to push the vacuum hoses off the throttle valve sensor fittings

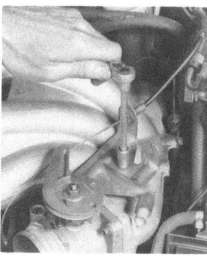

3.20b   After disconnecting the throttle cable, unbolt the cable bracket

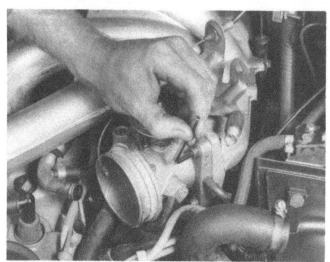

3.20c   Use an Allen wrench to remove the throttle valve assembly retaining bolts

3.20d   Lift the throttle valve up and unplug the throttle switch connector

and check terminal 9 or 3 of the connector to make sure that the specified voltage is being supplied to the sensor **(see illustration)**.

5   If the voltage with the connector unplugged is as specified, but the voltage with the connector plugged into the sensor is not, there is a fault in the air flow sensor and it should be replaced with a new one.

6   Check the air flow sensor door switch by pushing the sensor door open and holding it while checking the voltage (voltmeter on miniscale) at terminal 7 (non-turbocharged models) or terminal 2 (turbocharged models) and compare this to the Specifications **(see illustration)**.

7   To check the air flow sensor internal temperature sensor, connect a voltmeter between terminals 6 and 22 (the outer 2 pins) on the air flow sensor connector and compare the voltage reading to the Specifications **(see illustration)**.

8   If the air flow sensor fails any of the checks, replace it with a new one.

9   To remove the air flow sensor, unplug the electrical connector, unbolt the wiring harness bracket from the sensor and disconnect the air pipe from the sensor.

10 Remove the four retaining bolts and lift the sensor from the engine.

Engine temperature sensor

11 Unplug the connector from the control unit located under the instrument panel adjacent to the steering column **(see illustration)**.

12 Check between terminal 13 and ground with an ohmmeter and compare this reading to the Specifications.

13 If the readings are not as specified, unplug the temperature switch and repeat the check on the switch terminals **(see illustration)**.

14 Replace the switch with a new one if it fails to produce the specified ohmmeter reading.

15 To replace the temperature switch, unplug the connector and unscrew the sensor with a wrench.

Throttle switch

16 Unplug the control unit connector and check between a good ground and terminal 2 (non-turbocharged models) or terminal 12 (turbo-charged models) on the connector.

17 Have an assistant open the throttle all the way and record the ohmmeter reading. The ohmmeter should go to infinite ohms as the throttle is opened, even slightly.

18 Close the throttle fully and record the reading. The ohmmeter should go to zero.

19 On non-turbocharged models the switch can be further checked by connecting the ohmmeter between terminal 2 on the control unit connector and ground. With the throttle closed there should be a reading of infinite ohms. With the throttle open the reading should be zero.

20 If the readings are not as specified, remove the throttle switch for a further check as follows. Disconnect the air intake hose and vacuum hoses **(see illustration)**. Remove the throttle linkage and bracket **(see illustration)**, remove the Allen head retaining bolts, lift the assembly up, unplug the connector and lift off the switch assembly **(see illustrations)**.

21 Repeat the checks directly on the switch connector by connecting the ohmmeter probe to the center (ground) terminal and the terminal 3 on the switch **(see illustration)**. With the switch grounded and the probe on terminal 3 the readings should be — throttle open: 0 ohms, throttle closed: infinite ohms.

22 Now repeat the test between ground and terminal 2 **(see illustration)**. The reading should be just the reverse — throttle closed: 0 ohms, throttle open: infinite ohms.

23 If the readings are not as specified, the switch can be adjusted. Loosen the screws, open the throttle and rotate the switch until the it can be felt against the inner stop **(see illustration)**. Tighten the screws securely. Repeat the checks and if adjustment did not bring the readings to within specifications, replace the switch with a new one.

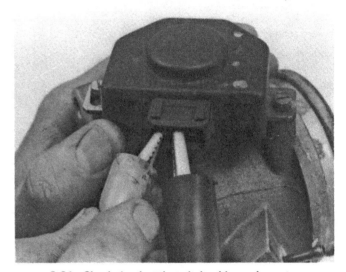

3.21 Check the throttle switch with an ohmmeter between the center (ground) and the number 3 terminals

3.22 Check the switch with an ohmmeter between the center and number 2 terminal. Note the terminal markings on top of the connector

3.23 With the throttle lever (A) open, adjust the throttle switch by rotating it in the direction shown (B) until it can be felt to bottom against the internal stop

3.24 Clean and check the throttle valve assembly O-ring (arrow) for damage before installing the assembly

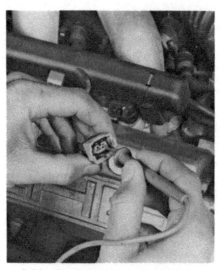

3.25 Check between a fuel injector terminal and a good ground to make sure battery voltage is reaching the injector

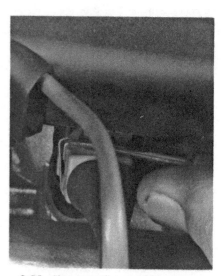

3.26 Use a small screwdriver to pry the connector spring retainer up so the connector can be unplugged

3.27 Unplug each fuel injector one at a time with the engine idling. If the speed does not change the injector is faulty

3.28 Check the resistance across the fuel injector terminal with an ohmmeter

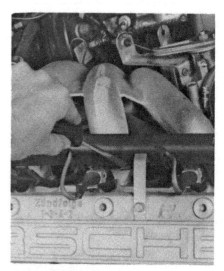

3.32 Release the fuel rail plastic cover clips with a screwdriver

24 Before installing the throttle switch, check the O-ring for damage (see illustration), replacing it with a new one if necessary.

Fuel injectors

25 With the engine Off and the ignition On, check between one of the injector connector terminals and ground to make sure there is battery voltage reaching the injector (see illustration). The injector ground is in the ECU, and one terminal of the injector connector should be hot whenever the ignition switch is On.
26 To check for an inoperative injector, release the connector wire retainer so the connector can be easily unplugged (see illustration).
27 With the engine idling, unplug each connector in turn (see illustration). If the engine speed does not change when an injector is unplugged, that injector is faulty and must be replaced.
28 Unplug the connector and check the resistance of the injector across the flat male plugs with an ohmmeter (see illustration).
29 Replace the injector with a new one if the resistance is not as specified.

Fuel rail removal and disassembly

30 Depressurize the fuel system (Section 2).
31 Disconnect the negative cable at the battery. Place the cable out of the way so it cannot accidentally come in contact with the negative terminal of the battery, as this would once again allow power into the electrical system of the vehicle.

Fuel rail

32 Remove the fuel rail cover (see illustration) and unplug all four injector connectors and move them out of the way.
33 Disconnect the regulator and damper vacuum connections.
34 Disconnect the fuel feed and return lines (see illustrations).
35 Disconnect the electrical harness from the rail (see illustration).
36 Remove the retaining bolts (see illustration).
37 Pull the rail straight up and out of the intake manifold openings (see illustration).

Pressure regulator and damper

38 Support the mount with pliers and unscrew the large fitting, followed by the small fitting (see illustrations)

Fuel injector

39 Pry out the clip with a small screwdriver, grasp the injector securely and work it out of the rail (see illustrations).

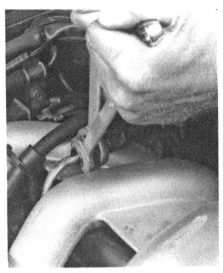

3.34a   Use two wrenches to disconnect the fuel feed line so the line won't be twisted

3.34b   Disconnect the fuel return hose at the pressure regulator

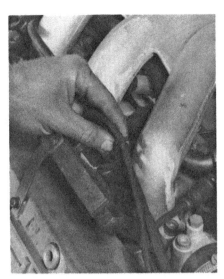

3.35   Detach the wiring harness tabs from the fuel rail

3.36   The fuel rail retaining bolts (arrows)

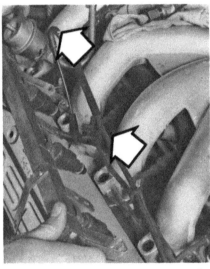

3.37   Pull the fuel rail/injector assembly straight out of the openings in the intake manifold

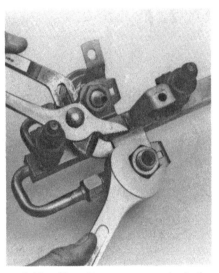

3.38a   Use pliers to support the fuel regulator or pressure damper mount while unscrewing the large nut

3.38b   Uscrew the smaller fitting using two wrenches

3.39a   Pry the fuel injector retaining clip off using a small screwdriver

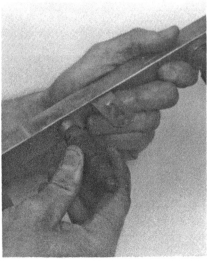

3.39b   Work the injector back and forth while pulling it out of the fuel rail

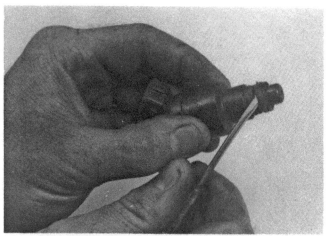

3.40 Pry carefully around the circumference of the O-rings to check for cuts or other damage

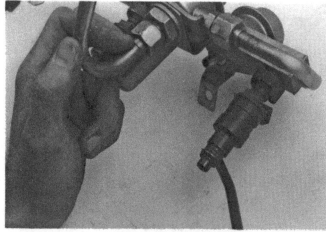

3.41 Apply automatic transmission fluid to the O-rings prior to installing the injectors

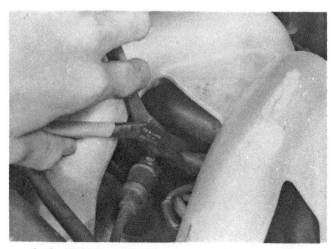

3.43 With the engine at normal operating temperature, the engine speed should drop when the auxiliary air hose is pinched shut

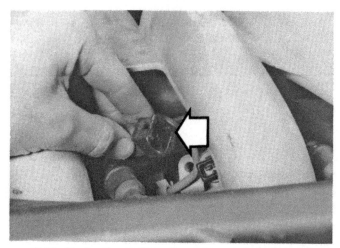

3.44 Unplug the auxiliary air regulator and check at the connector (arrow) to make sure there is battery voltage

40 Inspect the O-rings on either end of the injector for cuts or damage, replacing them with new ones if necessary (see illustration).
41 Coat the O-rings with clean automatic transmission fluid prior to installation (see illustration).
42 Installation of the fuel rail and components is the reverse of installation.

Auxiliary air regulator

43 Since the auxiliary air regulator will be closed when the engine is at normal operating temperature, pinching the air hose will cause the engine speed to drop when the engine is warmed up and idling (see illustration).
44 Unplug the connector and check the terminal voltage to verify that it is receiving battery voltage (see illustration).
45 Use an ohmmeter to check the regulator resistance. Replace regulator if the resistance is not within specifications.
46 To replace the regulator, remove the exhaust manifold (Chapter 2), remove the bolts and lift the regulator assembly from the engine.

---

4   Fuel tank — removal and installation

Warning: *Gasoline is extremely flammable, so extra precautions must be taken when working on any part of the fuel system. Do not smoke or allow open flames or bare light bulbs near the work area. Also, do not work in a garage if a natural gas-type appliance with a pilot light is present. While performing any work on the fuel tank it is advisable to have a fire extinguisher on hand and to wear safety glasses.*

1 Depressurize the fuel system (Section 2).
2 Disconnect the negative cable at the battery. Place the cable out of the way so it cannot accidentally come in contact with the negative terminal of the battery, as this would once again allow power into the electrical system of the vehicle.
3 Raise the vehicle and support it securely on jackstands.
4 Using a pump, remove all fuel from the tank.
5 Unbolt and remove the muffler and tail pipe assembly and heat shield.
6 Remove the fuel pump cover (if equipped), pinch the fuel line closed with a clamp or locking pliers and disconnect the fuel hoses and electrical connectors from the pump.
7 Remove the right side retaining strap from the fuel tank.
8 Remove the fuel pump.
9 Remove the transmission (Chapter 7).
10 Remove the fuel filter and mount.
11 Remove the floor cover and fuel level sender cover in the trunk compartment.
12 Disconnect the fuel return hose and unplug the sender electrical connector.
13 Remove the right side rear window.
14 Remove the inner trim panel and the insulation sheet covering the fuel tank filler pipe and vent hoses.
15 Loosen the fuel filler neck clamp and disconnect the vent hoses.
16 Under the vehicle, loosen the left side tank retaining strap, remove the two retaining nuts, lower the tank downward and to the rear and remove it from the vehicle.
17 Installation is the reverse of removal.

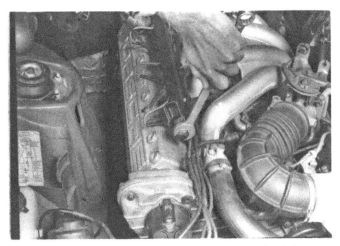

6.3   Unscrew the fuel rail cap with a wrench. Don't lose the sealing ball inside the cap

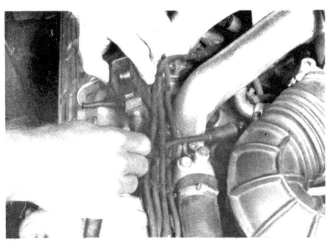

6.4   Push the hose tightly over the fuel rail threads

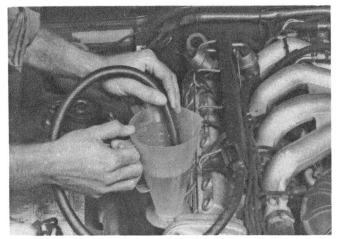

6.6   Run the engine for 30 seconds and then check how much fuel has run into the graduated container

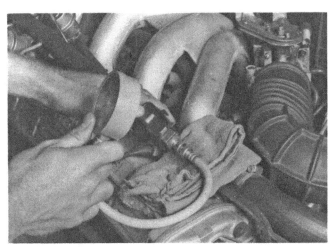

7.1   Place rags under the fuel rail so that no fuel leaks into the distributor when checking the fuel pressure

5   Fuel tank — cleaning and repair

1   Any repairs to the fuel tank or filler neck should be carried out by a professional who has experience in this critical and potentially dangerous work. Even after cleaning and flushing of the fuel system, explosive fumes can remain and ignite during repair of the tank.
2   If the fuel tank is removed from the vehicle, it should not be placed in an area where sparks or open flames could ignite the fumes coming out of the tank. Be especially careful inside garages where a natural gas-type appliance is located, because the pilot light could cause an explosion.

6   Fuel pump — checking

Refer to illustrations 6.3, 6.4 and 6.6
Warning: *Gasoline is extremely flammable, so extra precautions must be taken when working on any part of the fuel system. Do not smoke or allow open flames or bare light bulbs near the work area. Also, do not work in a garage if a natural gas-type appliance with a pilot light is present.*

1   Depressurize the fuel system (Section 2).
2   Disconnect the ignition coil positive (+) lead.
3   Unscrew the cap from the forward end of the fuel rail, taking care not to lose the sealing ball **(see illustration)**. **Warning:** *Place a rag under the opening to catch the residual fuel. Do not allow any fuel to drain into the distributor as this could cause an explosion when the engine*

is started.
4   To check the fuel pump output, you will need a length of suitable size hose to connect between the fuel rail and a graduated container with a capacity of approximately two liters. Push the hose tightly over the threads of the fuel opening and place the other end in the container **(see illustration)**.
5   Insert a jumper wire between terminals 30 and 87b in the fuse block located under the left side of the instrument panel. Refer to the underside of the fuel pump relay for the terminal markings.
6   Have an assistant turn the key On and allow the fuel pump to run for 30 seconds and then turn off the key or pull the jumper out **(see illustration)**. Check the amount of fuel in the container and compare this to the Specifications.
7   Replace the fuel pump with a new one if the output is significantly below the specified amount.

7   Fuel pressure test

Refer to illustrations 7.1, 7.3 and 7.4
1   Remove the plug from the end of the fuel rail, taking care not to lose the sealing ball. Connect a fuel pressure gauge to the fuel rail. It may be necessary to use a 11 x 1.50 mm adapter between the gauge and rail threads. Place a rag under the opening to catch the residual fuel. Do not allow any fuel to drain into the distributor as this could cause an explosion when the engine is started **(see illustration)**.
2   Start the engine and check the fuel pressure at idle. Compare this to the Specifications.

7.3   Pulling the fuel regulator vacuum hose off will cause
the fuel pressure to go up

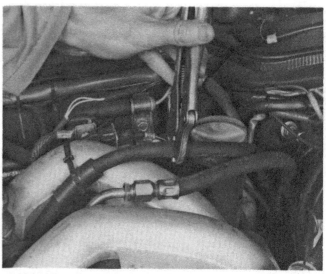

7.4   Clamping the fuel return hose shut will make the fuel
pressure rise to maximum

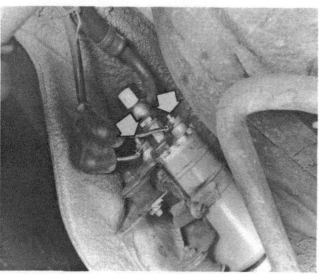

8.5   After pulling back the rubber covers disconnect the
fuel pump electrical leads (arrows)

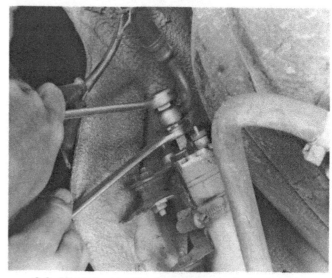

8.6   Use two wrenches to unscrew the fuel pump hose
union bolt

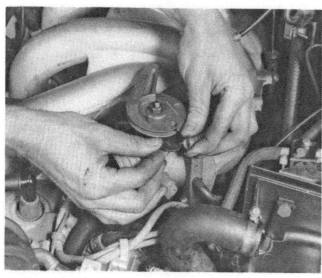

9.2   After disengaging the throttle cable, disconnect the
return spring

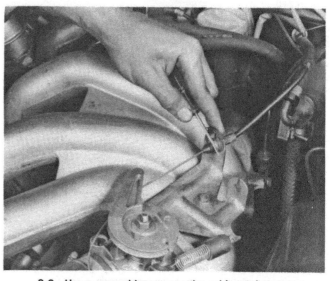

9.3   Use a screwdriver to pry the cable retainer out
of the mount

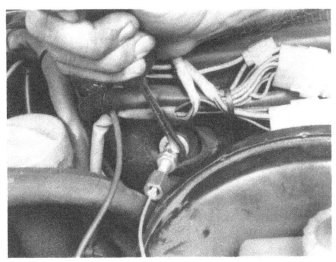

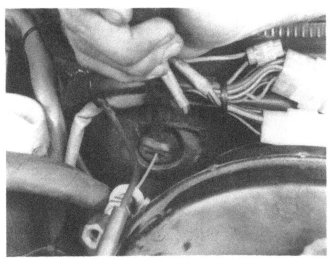

9.4a Pry the throttle cable guide out of the firewall grommet . . .

9.4b . . . followed by the grommet

3   Pull off the fuel regulator vacuum hose (see illustration). The fuel pressure should rise to between 33 and 39 psi.
4   If the pressure is not as specified, clamp the fuel return hose shut slowly (see illustration).
5   If the pressure rises significantly above 58 psi, the pressure regulator is faulty and it should be replaced with a new one.
6   If the fuel pressure is considerably below 58 psi there is a fault in the fuel pump or filter.

8   Fuel pump — removal and installation

Refer to illustrations 8.5 and 8.6
Warning: *Gasoline is extremely flammable, so extra precautions must be taken when working on any part of the fuel system. Do not smoke or allow open flames or bare light bulbs near the work area. Also, do not work in a garage if a natural gas type appliance with a pilot light is present.*

1   Depressurize the fuel system (Section 2).
2   Disconnect the negative cable at the battery. Place the cable out of the way so it cannot accidentally come in contact with the negative terminal of the battery, as this would once again allow power into the electrical system of the vehicle.
3   On later models, remove the fuel pump cover.
4   Lightly clamp the fuel hose closed between the tank and the pump using a suitable tool such as locking pliers.
5   Pull back the covers and disconnect the electrical leads from the pump (see illustration).
6   Disconnect the fuel hose clamp and pull the rear hose off the fuel pump. Use two wrenches to unscrew the fuel union bolt and disconnect the front fuel hose (see illustration).
7   Loosen the clamp screw and slide the fuel pump out of the mounting clamp.
8   Installation is the reverse of removal.

9   Throttle cable — removal, installation and adjustment

Refer to illustrations 9.2, 9.3, 9.4a and 9.4b
1   In the passenger compartment, pry the cable off the throttle pedal arm.
2   In the engine compartment, disconnect the cable from the throttle lever by pushing the lever all the way to the open position and then over center until the cable end can be disengaged. Disconnect the throttle return spring (see illustration).
3   Pry the cable retainer out of the intake manifold mount with a screwdriver and disengage the cable (see illustration).
4   Use a screwdriver to pry first the retainer and then the grommet out of the firewall (see illustrations).

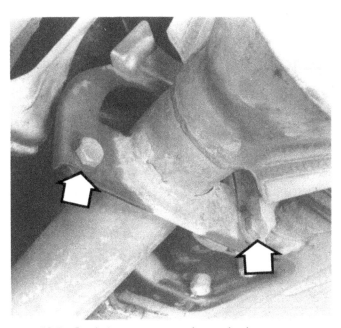

10.5   Catalytic converter-to-exhaust pipe hanger nuts and bolts (arrows)

5   Pull the throttle cable through into the engine compartment and remove it from the vehicle.
6   Installation is the reverse of removal.

10   Exhaust system components — removal and installation

Refer to illustrations 10.5, 10.6 and 10.7
Caution: *The exhaust system generates high temperatures. No part of the exhaust should be touched until the entire system has completely cooled. Be especially careful around the catalytic converter, where the highest temperatures are generated.*

1   Disconnect the negative cable at the battery. Place the cable out of the way so it cannot accidentally come in contact with the negative terminal of the battery, as this would once again allow power into the electrical system of the vehicle.
2   Raise the vehicle and support it securely on jackstands.
3   Remove the oxygen sensor (Chapter 6).
4   Remove the six flange nuts and bolts and disconnect the exhaust pipe from the exhaust manifold flanges.
5   Disconnect the catalytic converter hanger bolts (see illustration).

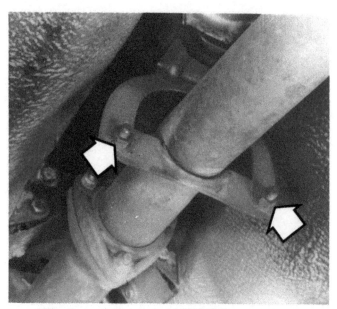

10.6   Rear exhaust pipe and catalytic converter hanger
nuts and bolts (arrows)

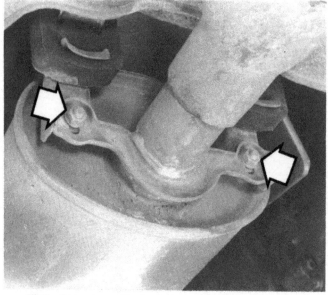

10.7   The rear muffler hanger nuts (arrows) support the
rear of the exhaust system

6   Unbolt and disconnect the rear exhaust pipe hanger **(see illustration)**.
7   Remove the two muffler-to-rubber hanger nuts **(see illustration)**
and lower the exhaust system from the vehicle.
8   Installation is the reverse of removal.

11   Turbocharger — general information

The turbocharger system increases power by using an exhaust gas
driven turbine to pressurize the air/fuel mixture as it enters the combustion chamber. The amount of boost (intake manifold pressure) is
controlled by the wastegate (exhaust bypass valve). This is operated
by a spring-loaded actuator assembly which controls the maximum
boost level by allowing some of the exhaust gas to bypass the turbine.
While a comparatively simple design, the turbocharger is a precision
device which can be severely damaged by an interrupted oil or coolant
supply or loose or damaged ducting.

Due to the special techniques and equipment required, any checking
or diagnosis of suspected problems should be left to your dealer. The
home mechanic can, however, check the connections and linkages for
security, damage or obvious faults.
Because each turbocharger has its own distinctive sound, a change
in the noise level can be a sign of potential problems. A high pitched
or whistling sound is a symptom of an inlet or exhaust gas leak. If an
unusual sound issues from the vicinity of the turbine, have it checked
by a dealer or properly equipped shop.
Inspect the exhaust manifold periodically for cracks and loose connections.
The turbine wheel rotates at speeds up to 140,000 rpm, so severe
damage can result from the interruption of coolant or contamination
of the oil supply to the turbine bearings. Check for leaks in the coolant
and oil inlet lines. Burned oil on the turbine housing is a sign of this.
Caution: *Any time a major engine bearing such as a main, connecting
rod or camshaft bearing is replaced, the turbocharger should be flushed
with clean oil.*

Chapter 5 Engine electrical systems

Contents

Specifications

Cylinder numbers (front to rear)	1–2–3–4
Firing order	1–3–4–2
Ignition coil resistance	
Primary	0.4 to 0.6 ohms
Secondary	5 to 7.2 K ohms

1 General information

Warning: *Because of the very high voltage generated by the DEE system, extreme care should be taken whenever an operation involving ignition components is performed. This not only includes the distributor, coil, control module and spark plug wires, but related items such as the plug connections, tachometer and testing equipment.*

The engine electrical system includes the battery, charging system, starter and the Digital Engine Electronic (DEE) ignition and electronic fuel injection system.

The DEE ignition system consists of the control unit, the battery, the coil, the primary (low tension) and secondary (high tension) wiring circuits, the distributor and the spark plugs. All spark timing changes in the DEE system are carried out electronically by the control unit, which monitors data from various engine sensors, computes the desired spark timing and changes the timing accordingly.

The electronic fuel injection consists of the electric fuel pump, the fuel injectors, the air flow sensor, the throttle switch and the control unit. Information on the electronic fuel injection system can be found in Chapter 4.

The charging system is made up of the alternator, with an integral regulator incorporating the brush holder assembly, and the battery. The starter is operated by the battery's electrical power.

Information on the routine maintenance of the ignition, starting and charging systems and the battery can be found in Chapter 1.

2 Battery — removal and installation

Refer to illustration 2.1

Warning: *Hydrogen gas is produced by the battery, so keep open flames*

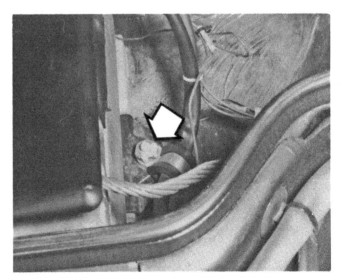

2.1 The battery clamp nut (arrow) is located at the base of the battery — keep it tight to avoid damage to the battery

and lighted cigarettes away from it at all times. Always wear eye protection when working around a battery. Rinse off spilled electrolyte immediately with large amounts of water.

1 The battery is located at the right rear corner of the engine compartment and is held in place by a hold-down clamp at its base **(see illustration)**.

2   Lift off the battery cover and disconnect the battery cables. Always disconnect the negative battery cable first, followed by the positive cable.
3   After the cables have been disconnected, remove the nut and hold-down clamp.
4   Remove the battery. **Warning:** *When lifting the battery from the engine compartment, be careful not to twist the case as acid could spurt out of the filler openings.*
5   Installation is the reverse of removal. Make sure the battery holder is clean before setting the battery in it.

3   Battery — emergency jump starting

Refer to the *Booster battery (jump) starting* procedure at the front of this manual.

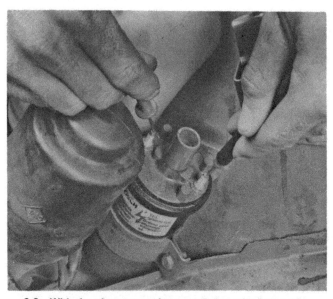

6.2   With the ohmmeter selector switch on the Low scale, check the resistance between terminals 1 and 15 on the ignition coil

4   Battery cables — check and replacement

1   Periodically inspect the entire length of each battery cable for damage, cracked or burned insulation and corrosion. Poor battery cable connections can cause starting problems and decreased engine performance.
2   Check the cable-to-terminal connections at the ends of the cables for cracks, loose wire strands and corrosion. The presence of white, fluffy deposits under the insulation at the cable terminal connection is a sign that the cable is corroded and should be replaced. Check the terminals for distortion, missing mounting bolts and corrosion.
3   If only the positive cable is to be replaced, be sure to disconnect the negative cable from the battery first.
4   Disconnect and remove the cable. Make sure that the replacement cable is the same length and diameter.
5   Clean the threads of the starter or ground connection with a wire brush to remove rust and corrosion. Apply a light coat of petroleum jelly to the threads to ease installation and prevent future corrosion.
6   Attach the cable to the starter or ground connection and tighten the mounting nut securely.
7   Before connecting the new cable to the battery, make sure that it reaches the terminal without having to be stretched.
8   Connect the positive cable first, followed by the negative cable.

5   Distributor — removal and installation

These models do not have a conventional distributor. The distributor rotor is located on the front of the camshaft, which extends through the camshaft cover. The distributor cap mounts over the rotor on the camshaft cover. Only the cap and rotor can be removed — this procedure is covered in Section 26 of Chapter 1.

6   Ignition coil — check and replacement

Refer to illustrations 6.2, 6.3a and 6.3b
Warning: *The very high voltage generated by the ignition system can transmit a possibly fatal shock when the engine is running. Consequently, extreme care should be taken whenever an operation is performed involving ignition components. The ignition switch should be*

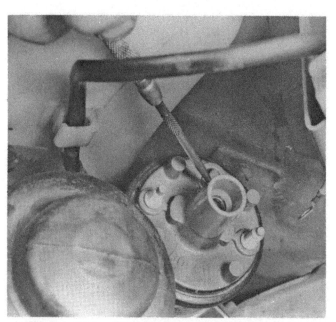

6.3a   Terminal 4 is located under a small rubber plug in the coil case — pry the plug out to expose the terminal, . . .

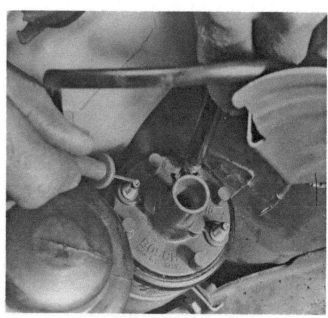

6.3b   . . . then take the resistance reading with the ohmmeter selector switch on the High scale

turned off or the battery ground cable should be disconnected when doing any work on the ignition system.

1   Remove the cover and disconnect the primary and secondary wires from the coil. The primary wires are attached to the terminals with nuts — the secondary wire (to the distributor cap) can be pulled out.
2   To check the primary resistance, connect an ohmmeter between terminals 1 and 15 **(see illustration)**. Set the ohmmeter selector switch to the Low scale. Compare the ohmmeter reading to the Specifications.
3   To check the secondary resistance, set the ohmmeter selector switch to the High scale, then connect the leads to terminals 1 and 4 **(see illustrations)**. Compare this reading to the Specifications.
4   To remove the coil, loosen the clamp bolt and withdraw it from the mounting bracket.
5   Installation is the reverse of removal.

7   Charging system — general information and precautions

The charging system includes the alternator, voltage regulator and battery. These components work together to supply electrical power for the ignition system, lights, radio, etc. The alternator is driven by a drivebelt at the front of the engine.

The purpose of the voltage regulator is to limit the alternator's voltage to a preset value. This prevents power surges, circuit overloads, etc., during peak voltage output. On all models with which this manual is concerned, the voltage regulator is part of the brush holder assembly.

The charging system does not ordinarily require periodic maintenance. The drivebelt, electrical wiring and connections should, however, be inspected at the intervals suggested in Chapter 1.

Take extreme care when making circuit connections to a vehicle equipped with an alternator and note the following. When making connections to the alternator from a battery, always match correct polarity. Before using arc welding equipment to repair any part of the vehicle, disconnect the wires from the alternator and the battery terminals. Never start the engine with a battery charger connected. Always disconnect both battery leads before using a battery charger.

8   Charging system — check

Refer to illustration 8.4

1   If a malfunction occurs in the charging circuit, do not immediately

assume that the alternator is causing the problem. First check the following items:
a)   The battery cables where they connect to the battery. Make sure the connections are clean and tight.
b)   Check the external alternator wiring and connections. They must be in good condition.
c)   Check the drivebelt condition and tension (Chapter 1).
d)   Make sure the alternator mounting bolts are tight.
e)   Run the engine and check the alternator for abnormal noise.
2   Using a voltmeter, check the battery voltage with the engine off. It should be approximately 12-volts.
3   Start the engine and check the battery voltage again. It should now be approximately 14 to 15-volts.
4   Locate the test hole in the back of the alternator and ground the tab that is located inside the hole by inserting a screwdriver blade into the hole and touching the tab and the case at the same time **(see illustration)**. **Caution:** *Do not run the engine with the tab grounded any longer than is necessary to obtain a voltmeter reading. If the alternator is charging, it is running unregulated during the test. This condition may overload the electrical system and cause damage to the components.*
5   The reading on the voltmeter should be 15-volts or higher with the tab grounded in the test hole.
6   If the voltmeter indicates low battery voltage, the alternator is faulty and should be replaced with a new one (Section 9).
7   If the voltage reading is 15-volts or higher and a no charge condition is present, the regulator or field circuit is the problem. Remove the alternator and have it checked further by an auto electric shop.

9   Alternator — removal and installation

Refer to illustrations 9.4, 9.5 and 9.6

1   Disconnect the negative cable at the battery. Place the cable out of the way so it cannot accidentally come in contact with the negative terminal of the battery, as this would once again allow power into the electrical system of the vehicle.
2   Remove the air cleaner assembly.
3   Under the vehicle, loosen and remove the accessory drivebelt and unbolt the adjuster from the alternator.
4   In the engine compartment, disconnect the two wires from the alternator **(see illustration)**.

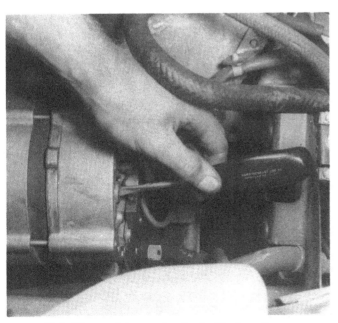

8.4   The regulator can be checked by inserting a screwdriver into the test hole and grounding the tab against the case while taking a voltmeter reading

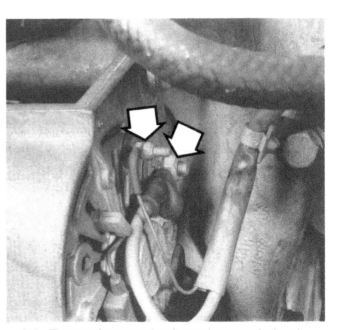

9.4   The two alternator wires (arrows) are attached to the terminals with nuts

5   Remove the through-bolt and nut **(see illustration)**.
6   On models so equipped, remove the alternator-to-air conditioning compressor bolt **(see illustration)**.
7   Lift the alternator from the vehicle.
8   Installation is the reverse of removal. Refer to Chapter 1 and adjust the drivebelt.

10   Alternator brush/regulator assembly — replacement

Refer to illustrations 10.2 and 10.3

1   Disconnect the negative cable at the battery. Place the cable out of the way so it cannot accidentally come in contact with the negative terminal of the battery, as this would once again allow power into the electrical system of the vehicle.
2   Remove the brush holder/regulator assembly mounting screws **(see illustration)**.

3   Separate the assembly from the rear of the alternator and tilt it while lifting it up and off the alternator **(see illustration)**.
4   Installation is the reverse of removal.

11   Starting system — general information

The function of the starting system is to crank the engine. The starting system is composed of a starting motor, solenoid and battery. The battery supplies the electrical energy to the solenoid, which then completes the circuit to the starting motor, which does the actual work of cranking the engine.

The electrical circuitry of the vehicle is arranged on automatic transmission models so that the starter motor can only be operated when the transmission selector lever is in Park or Neutral.

Never operate the starter motor for more than 30 seconds at a time

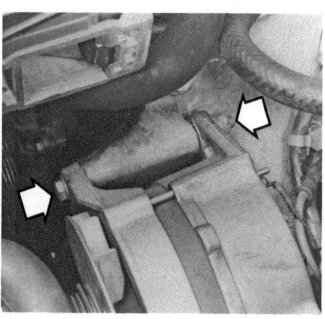

9.5   The alternator through-bolt and nut (arrows)

9.6   Alternator-to-air conditioning compressor bolt

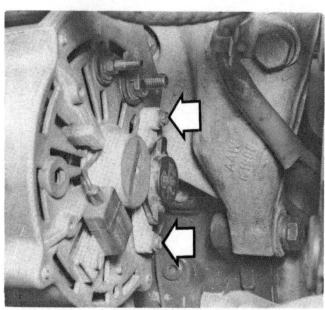

10.2   The brush holder and regulator is attached to the alternator by two screws (arrows)

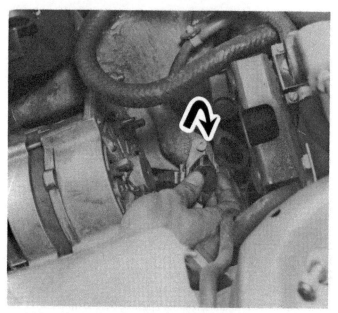

10.3   Tilt the brush holder and regulator assembly out of the alternator

without pausing to allow it to cool for at least two minutes. Excessive cranking can cause overheating, which can seriously damage the starter.

12 Starter motor — testing in vehicle

Refer to illustration 12.6

Note: *Before diagnosing starter problems, make sure that the battery is fully charged.*

1   If the starter motor does not turn at all when the switch is operated, make sure that the shift lever is in Park (automatic transmission).
2   Make sure that the battery is charged and that all cables, both at the battery and starter solenoid terminals, are secure.
3   If the starter motor spins but the engine is not cranking, then the overrunning clutch in the starter motor is slipping and the starter motor must be removed from the engine and disassembled.

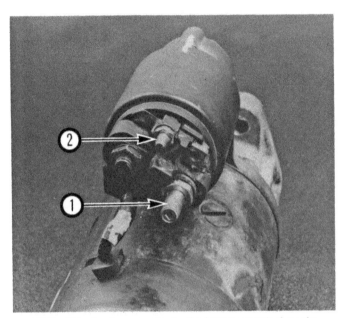

12.6   Attach a jumper wire to the battery terminal on the solenoid (1) (or the positive battery post) and the switch terminal (2) to bypass the ignition switch

4   If, when the switch is actuated, the starter motor does not operate at all but the solenoid clicks, then the problem lies with either the battery, the main solenoid contacts or the starter motor itself.
5   If the solenoid plunger cannot be heard when the switch is actuated, the solenoid itself is defective or the solenoid circuit is open.
6   To check the solenoid, connect a jumper lead between the positive terminal of the battery and the switch terminal on the solenoid (the switch terminal is the small one at the top) **(see illustration)**. If the starter motor now operates, the solenoid is OK and the problem is in the ignition switch, Neutral start switch or the wiring.
7   If the starter motor still does not operate, remove the starter/solenoid assembly for disassembly, testing and repair.
8   If the starter motor cranks the engine at an abnormally slow speed, first make sure that the battery is charged and that all terminal connections are tight. If the engine is partially seized, or has the wrong viscosity oil in it, it will crank slowly.
9   Run the engine until normal operating temperature is reached, then disconnect the coil wire from the distributor cap and ground it on the engine.
10   Connect a voltmeter positive lead to the starter motor terminal of the solenoid and then connect the negative lead to ground.
11   Actuate the ignition switch and take the voltmeter readings as soon as a steady figure is indicated. Do not allow the starter motor to turn for more than 30 seconds at a time. A reading of 9-volts or more, with the starter motor turning at normal cranking speed, is normal. If the reading is 9-volts or more but the cranking speed is slow, the motor is faulty. If the reading is less than 9-volts and the cranking speed is slow, the solenoid contacts are probably burned.

13 Starter motor — removal and installation

Refer to illustrations 13.3 and 13.4

1   Disconnect the negative cable at the battery. Place the cable out of the way so it cannot accidentally come in contact with the negative terminal of the battery, as this would once again allow power into the electrical system of the vehicle.
2   Raise the vehicle and support it securely on jackstands.
3   Disconnect the wires from the starter motor solenoid. Note that the upper terminal (for the battery cable and electrical system hot wire) is larger in diameter than the terminal to the right (for the ignition switch wire), so it is impossible to confuse the wires when reattaching them to the solenoid terminals **(see illustration)**.
4   Remove the mounting bolts, separate the starter motor from the bellhousing and lower it from the vehicle **(see illustration)**.
5   Installation is the reverse of removal.

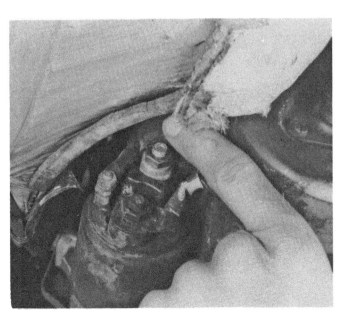

13.3   The starter electrical connections are different sizes, preventing confusion during installation

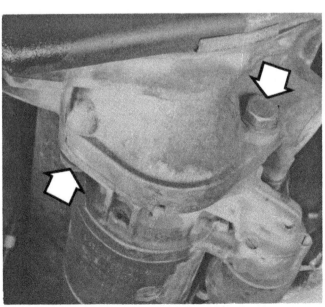

13.4   Starter motor mounting bolts (arrows)

14.2a   Remove the nut and detach the starter solenoid field coil cable . . .

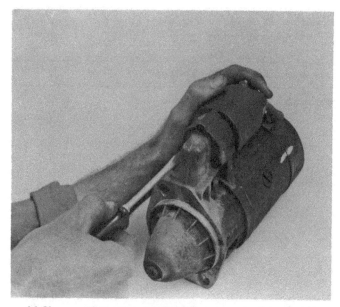

14.2b   . . . then use a screwdriver to remove the solenoid mounting screws

14.3   Mark the relationship of the solenoid and starter motor housing so they can be reinstalled in the same relative position

14.4   The spring and plunger can fall out, so make sure they are reinstalled as shown

14   Solenoid — replacement

Refer to illustrations 14.2a, 14.2b, 14.3 and 14.4

1   Remove the starter motor (Section 13).
2   Disconnect the field coil cable and remove the three mounting screws **(see illustrations)**.
3   Mark the solenoid housing-to-starter relationship **(see illustration)**.
4   Detach the solenoid from the starter. The plunger and spring may fall out of the solenoid, so make sure the components are properly assembled prior to installation **(see illustration)**.
5   Installation is the reverse of removal.

Chapter 6   Emissions control systems

Contents

Specifications

Torque specifications	Ft-lbs	Nm
Oxygen sensor	30	40

1   General information

Refer to illustration 1.6

To prevent pollution of the atmosphere from incompletely burned and evaporating gases, and to maintain good driveability and fuel economy, a number of emission control devices are incorporated. They include the:

Electronic fuel injection system
Evaporative Emission Control System (EECS)
Catalytic converter

These systems are linked, directly or indirectly, to the Digital Engine Electronics (DEE) control system (Chapters 4 and 5).

The Sections in this Chapter include general descriptions, checking procedures within the scope of the home mechanic and component replacement procedures (when possible) for each of the systems listed above.

Before assuming that an emissions control system is malfunctioning, check the fuel and ignition systems carefully. The diagnosis of some emission control devices requires specialized tools, equipment and training. If checking and servicing become too difficult or if a procedure is beyond the scope of your skills, consult your dealer service department.

This doesn't mean, however, that emission control systems are particularly difficult to maintain and repair. You can quickly and easily perform many checks and do most (if not all) of the regular maintenance at home with common tune-up and hand tools. **Note:** *The most frequent cause of emissions problems is simply a loose or broken vacuum hose or wiring connection, so always check the hose and wiring connections first.*

Pay close attention to any special precautions outlined in this Chapter. It should be noted that the illustrations of the various systems may not exactly match the system installed on your vehicle because of changes made by the manufacturer during production or from year to year.

A *Vehicle Emissions Control Information* label is located in the engine compartment **(see illustration)**. This label contains important emissions specifications and setting procedures, as well as a vacuum hose schematic with emissions components identified. When servicing the engine or emissions systems, the VECI label in your particular vehicle should always be checked for up-to-date information.

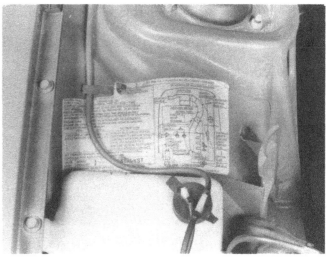

1.6   The Vehicle Emission Control Information (VECI) label provides information regarding engine size, exhaust emission system used, engine adjustment procedures and specifications and an emission component and vacuum line schematic diagram

2   Digital Engine Electronics (DEE) system

The Digital Engine Electronics (DEE) system consists of a control unit and information sensors (notably the oxygen sensor in the exhaust system) which monitor various functions of the engine and send data back to the control unit. The control unit in turn sends out signals to the ignition and fuel injection systems in accordance with driving conditions.

The DEE system sensors constantly relay information to the control unit, which processes the data and, if necessary, sends out a command to change the operating parameters of the engine.

Here's a specific example of how one portion of this system operates: An oxygen sensor, located in the exhaust pipe, constantly monitors the oxygen content of the exhaust gas. If the percentage of oxygen in the exhaust gas is incorrect, an electrical signal is sent to the control unit. The control unit takes this information, processes it and then sends a command to the fuel injection system, telling it to change the air/fuel mixture. This happens in a fraction of a second and it goes on continuously when the engine is running. The end result is an air/fuel mixture ratio which is constantly maintained at a predetermined ratio, regardless of driving conditions.

One might think that a system which uses an on-board computer and electrical sensors would be difficult to diagnose. This is not necessarily the case. The following DEE system components directly involved with the emissions system can be individually checked and replaced, referring to the appropriate chapter. If the problem persists after these checks have been made, more detailed service procedures will have to be done by a dealer service department.

> *Fuel injection components (Chapter 4)*
> *Air flow sensor*
> *Throttle switch*
> *Engine temperature sensor*
> *Fuel injectors*
> *Auxiliary air regulator*
> *Pressure regulator and damper*
> *Evaporative Emissions Control System (EECS) (Section 5)*

---

3   Control unit — removal and installation

Refer to illustrations 3.3, 3.4a and 3.4b

1   The Digital Engine Electronics control unit is located under the instrument panel, adjacent to the left side of the steering column.
2   Disconnect the negative cable at the battery. Place the cable out of the way so it cannot accidentally come in contact with the negative terminal of the battery, as this would once again allow power into the electrical system of the vehicle.
3   Release the clip at the cable end, unplug the connector and then swing the connector (which pivots at the opposite end) out of the unit **(see illustration)**.
4   Remove the mounting bolts and lower the control unit from the vehicle **(see illustrations)**.
5   Installation is the reverse of removal.

---

4   Oxygen sensor

Refer to illustration 4.6

General description

1   The oxygen sensor, which is located on the top side of the exhaust pipe just below the joint between the manifold and the pipe, monitors the oxygen content of the exhaust gas stream. The oxygen content in the exhaust reacts with the oxygen sensor to produce a voltage output. The control unit monitors this voltage output to determine the ratio of oxygen to fuel in the mixture. The control unit alters the air/fuel mixture ratio by controlling the pulse width (open time) of the fuel injec-

tors, thus allowing the catalytic converter to operate at maximum efficiency. It is this ratio which the control unit and the oxygen sensor attempt to maintain at all times.
2   The proper operation of the oxygen sensor depends on three conditions:

   a) **Electrical** — The low voltages and low currents generated by the sensor depend upon good, clean connections which should be checked whenever a malfunction of the sensor is suspected or indicated.
   b) **Proper operating temperature** — The control unit will not react to the sensor signal until the sensor reaches normal operating temperature for the engine. This factor must be taken into consideration when evaluating the performance of the sensor.
   c) **Unleaded fuel** — The use of unleaded fuel is essential for proper operation of the sensor. Make sure the fuel you are using is of this type.

3   In addition to observing the above conditions, special care must be taken whenever the sensor is serviced.

   a) The oxygen sensor has a permanently attached pigtail and connector which should not be removed from the sensor. Damage or removal of the pigtail or connector can adversely affect operation of the sensor.
   b) Grease, dirt and other contaminants should be kept away from the electrical connector and the end of the sensor.
   c) Do not use cleaning solvents of any kind on the oxygen sensor.
   d) Do not drop or roughly handle the sensor.

Replacement

Note: *Because the oxygen sensor is located in the exhaust pipe, it may be too tight to remove when the engine is cold. If you find it difficult*

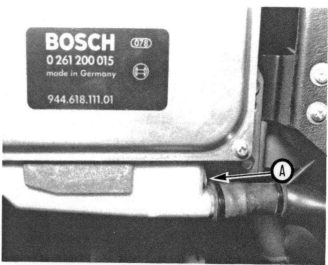

3.3   Release the control unit connector clip (A) and swing the connector out of the unit

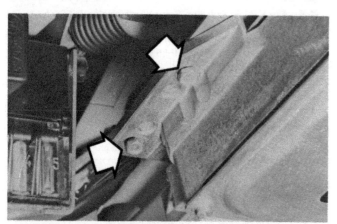

3.4a   The control unit is held in place by two bolts on the left . . .

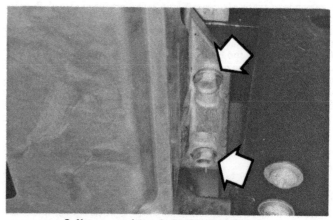

3.4b   . . . and two bolts on the right side

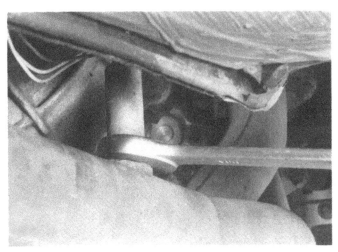

4.6   Use a wrench to unscrew the oxygen sensor from the exhaust pipe

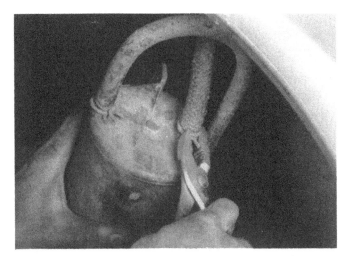

5.10   Use a pair of pliers to move the clamp up and off the fitting, then disconnect the hose from the canister

to loosen, start and run the engine for a minute or two, then shut it off. Be careful not to burn yourself during the following procedure.

4   Disconnect the negative cable at the battery. Place the cable out of the way so it cannot accidentally come in contact with the negative terminal of the battery, as this would once again allow power into the electrical system of the vehicle.

5   Carefully unplug the sensor lead from the engine wiring harness.

6   Raise the vehicle and support it securely on jackstands. Unscrew the sensor from the exhaust manifold **(see illustration)**. **Caution:** *Excessive force may damage the threads.*

7   Anti-seize compound must be used on the threads of the sensor to facilitate future removal. The threads of new sensors will already be coated with this compound, but if an old sensor is removed and reinstalled, recoat the threads.

8   Install the sensor and tighten it to the specified torque.

9   Reconnect the sensor lead to the main engine wiring harness.

10   Lower the vehicle and reconnect the cable to the negative terminal of the battery.

5   Evaporative Emission Control System (EECS)

Refer to illustration 5.10

General description

1   This system is designed to trap and store fuel vapors that evaporate from the fuel tank, fuel system and intake manifold.

2   The Evaporative Emission Control System (EECS) consists of a charcoal filled canister (located in the left front wheel well) and the lines connecting the canister to the fuel tank and fuel system.

3   Fuel vapors are transferred from the fuel tank and fuel injection system to the canister, where they are stored when the engine is not operating. When the engine is running, the fuel vapors are purged from the canister by intake air flow and consumed in the normal combustion process.

4   Evidence of fuel loss or a strong fuel odor are symptoms of a damaged or faulty evaporative system.

Checking

5   Inspect each hose attached to the canister for kinks, leaks and breaks along its entire length. Repair or replace as necessary.

6   Inspect the canister. It it's cracked or damaged, replace it.

7   Look for fuel leaking from the bottom of the canister. If fuel is leaking, replace the canister. Check the hoses and hose routing.

Component replacement

8   Raise the front of the vehicle, support it securely on jackstands and remove the left front wheel.

9   Remove the mounting bolts and lower the canister.

10 Disconnect the hose from the canister, using a pair of pliers to slide off the clamp **(see illustration)**.

11 Remove the canister.

12 Installation is the reverse of removal.

6   Catalytic converter

General description

1   The catalytic converter is an emission control device added to the exhaust system to reduce pollutants from the exhaust gas stream.

Checking

2   The test equipment for a catalytic converter is expensive and highly sophisticated. If you suspect that the converter on your vehicle is malfunctioning, take it to a dealer or authorized emissions inspection facility for diagnosis and repair.

3   The converter is located underneath the right side of the passenger compartment. Therefore, whenever the vehicle is raised for servicing of underbody components, check the converter for leaks, corrosion and other damage.

Component replacement

4   Because the catalytic converter is welded into the exhaust system, replacement should be left to a dealer or properly equipped shop.

Chapter 7 Part A Manual transmission

Contents

Specifications

Torque specifications	Ft-lbs	Nm
Backup light switch	22	30
Central tube-to-transmission bolt		
12 mm	61	85
10 mm	30	42
Central tube-to-clutch housing bolt	30	42
Driveaxle bolt	30	42
Driveshaft-to-transmission input		
shaft clamp bolt	58	80
Shift rod-to-transmission case bolt	10	14
Shift rod-to-intermediate lever	15	21
Shift linkage selector rod bolt	15	21
Shifter bracket bolt	15	21
Shift lever plate-to-central tube	15	21
Intermediate shift lever-to-selector shaft	16	22
Transmission drain plug	18	25
Transmission fill plug	18	25
Transmission case-to-central tube bolt	61	85
Transmission mount-to-body bolt	33	46

1 General information

All models are equipped with either a 5-speed manual or a 3-speed automatic transmission. Information on the manual transmission is included in this Part of Chapter 7. Information for the automatic transmission can be found in Part B of this Chapter.

The transmission used in these vehicles is essentially two separate units coupled together. They consist of a 5-speed transmission bolted to a final drive unit and differential assembly, sharing a common oil supply.

Due to the complexity, unavailability of replacement parts and the special tools necessary, internal repair procedures are not recommended for the home mechanic. The information contained within this manual will be limited to general diagnosis, external adjustments and removal and installation.

Depending on the expense involved in having a faulty transmission overhauled, it may be an advantage to consider replacing the unit with either a new or rebuilt one. Your local dealer or transmission shop should be able to supply you with information concerning cost, availability and exchange policy. Regardless of how you decide to remedy a faulty transmission problem, however, you can still save considerable expense by removing and installing the unit yourself.

2 Manual transmission shift linkage — adjustment

Refer to illustrations 2.3

1   The shift linkage is properly adjusted when the shift lever is vertical in the Neutral position. If the shift lever leans to the right or left, it can be adjusted at the shift rod intermediate lever on the transmission to return it to a vertical position.
2   Raise the vehicle and support it securely on jackstands.
3   Under the vehicle, loosen the shift rod intermediate lever bolt and nut **(see illustration)**. Move the bolt in the slot until the shift lever in the passenger compartment is vertical. Tighten the bolt and nut securely.

2.3 Loosen the shift rod intermediate rod nut (arrow) and move the bolt in the slot to adjust the shift lever angle

3.4 Remove the retaining bolt while keeping the flange from turning with a punch

3.5 Withdraw the axleshaft flange for access to the seal

3.6 Pull the old seal out of the bore with hooked tool

3 Differential axleshaft flange seal — replacement

Refer to illustrations 3.4, 3.5, 3.6 and 3.7

1   Raise the vehicle and support it securely on jackstands.
2   Drain the oil from the transmission (Chapter 1).
3   Remove the appropriate driveaxle.
4   Insert a punch through the flange into the bolt hole in the transmission to keep it from turning and remove the retaining bolt (**see illustration**).
5   Withdraw the axleshaft flange from the transmission (**see illustration**).
6   Pry the old seal out of the bore with a suitable hooked tool (**see illustration**).
7   Place the new seal in position and tap it evenly into the bore using a hammer and seal driver tool or suitable size socket (**see illustration**).
8   Install the flange and driveaxle and lower the vehicle.

3.7 Tap the seal evenly into the bore with seal driver or socket

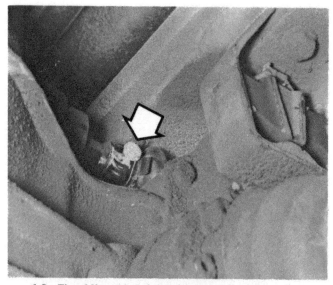

4.3   The shift rod bolt (arrow) is accessible after pulling the boot back

4.4   Grasp shift lever cover securely and pull up to disengage it

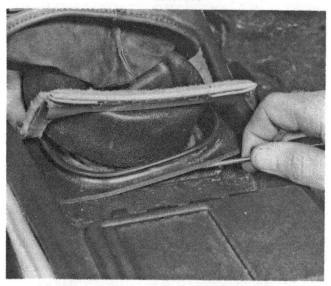

4.5   Use a small screwdriver to pry the trim clips out

4.6   Pull up on the shift boot

4.7a   Pry off the shift knob/boot cover retaining clip . . .

4   Manual transmission — removal and installation

Refer to illustrations 4.3, 4.4, 4.5, 4.6 4.7a, 4.7b, 4.8a, 4.8b, 4.9, 4.12, 4.13a, 4.13b, 4.14, 4.15, 4.16, 4.18 and 4.21

Removal

1   Disconnect the negative cable at the battery. Place the cable out of the way so it cannot accidently come in contact with the negative terminal of the battery, as this would once again allow power into the electrical system of the vehicle.

2   Raise the vehicle and support it securely on jackstands.

3   Under the vehicle, push back the shift rod boot at the top of the transmission, remove the retaining bolt and disconnect the rod (see illustration).

4   In the passenger compartment, disengage the leather cover from the shift boot and pull it up out of the way (see illustration).

5   Pull the carpet up and pry the shift console trim piece clips loose with a small screwdriver (see illustration).

6   Pull the shift boot off the shifter opening (see illustration).

7   Use a screwdriver to pry the clip off and remove the shift lever cover and knob assembly (see illustrations).

4.7b . . . and lift the assembly off

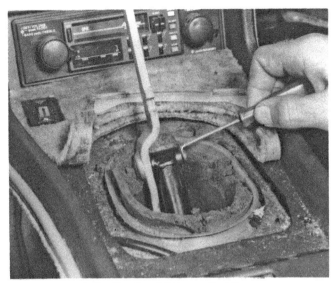

4.8a Pry the shift lever clip off . . .

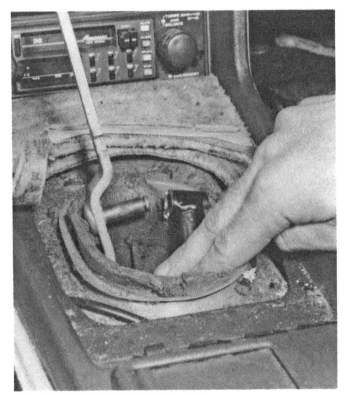

4.8b . . . and separate shift rod the from the lever

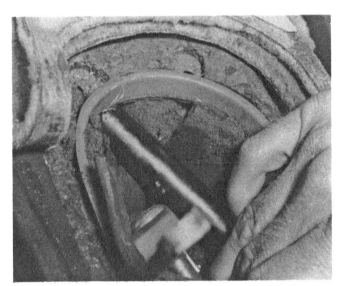

4.9 Turn the shift rod over for clearance and push
it forward

4.12 Mark the relationship of the driveshaft splines and
the coupler with white paint (arrow)

8   Remove the shift rod clip and separate the rod from the lever (see illustrations).
9   Turn the rod over and push it forward about 12 inches to pull it out of the transmission (see illustration).
10 Pry the rubber access hole covers out of the bottom of the transmission bellhousing.
11 With the help of an assistant, turn the rear wheels until the driveshaft retaining bolts are accessible through the inspection holes.
12 Mark the relationship of the driveshaft splines and the transmission couplers for reinstallation to the same position (see illustration).

4.13a   An Allen head tool and extension will be necessary
when removing the coupler bolts in the transmission . . .

4.13b   . . . and the central tube

4.14   Hang the driveaxle out of the way in a level position
so the CV joints won't be damaged

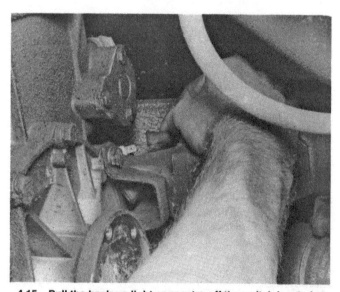

4.15   Pull the back-up light connector off the switch located at
the top of the transmission

4.16   The transmission-to-central tube bolt and nut
locations (arrows)

4.18   Transmission mount-to-chassis bolt location (arrow)

13 Use an Allen head tool and extension to unscrew the bolts from the couplers **(see illustrations)**.
14 Disconnect the driveaxles and hang them level so the CV joints won't be damaged **(see illustration)**.
15 Unplug the stop light switch **(see illustration)**.
16 Remove the transmission-to-central tube bellhousing nuts and bolts **(see illustration)**.
17 Support the transmission with a jack.
18 Remove the two transmission mount-to-chassis bolts **(see illustration)**.
19 Move the jack and transmission rearward and disconnect the driveshaft coupler.
20 Pull the shift levers back as far as they will go and use a hammer and a punch to disconnect the shift rod.
21 Lower the transmission and move it rearward while guiding the plastic shift rod tube out through the transmission hole to remove it from the housing **(see illustration)**.

Installation

22 Raise the transmission into place against the rear central tube bellhousing.
23 Insert the plastic shift tube into the transmission bellhousing.
24 Have an assistant push the shift rod back while you guide it into the shift lever on the transmission shift rod, making sure the dimple in the rod aligns with the hole in the shifter receptacle. Apply thread sealant on the threads of the lock bolt and install it. Tighten the bolt securely.
25 Install the transmission bellhousing bolts and nuts. Tighten them to the specified torque.
26 Working through the access hole, slide the coupling collars back over the transmission input splines. Tighten the coupler bolts securely.

4.21 Slide the shift rod tube out of the housing after separating the transmission from the central tube

27 Install the transmission mount-to-chassis bolts. Tighten the bolts to the specified torque.
28 The remainder of installation is the reverse of the removal procedure.

Chapter 7 Part B Automatic transmission

Contents

Specifications

Torque converter-to-transmission bellhousing clearance . . . . . 13/32 in (10 mm)

Torque specifications	Ft-lbs	Nm
Transmission bellhousing bolts		
10 mm	30	42
12 mm	61	85
Transmission mount bolts	30	42
Torque converter-to-driveplate bolts	29	40

1 General information

The transmission used in these vehicles is essentially two separate units coupled together; a 3-speed automatic transmission bolted to a final drive and differential unit, each having it's own fluid supply.

Information on the automatic transmission is included in this Part of Chapter 7. Information for the manual transmission can be found in Part A of this Chapter.

Due to the complexity of the clutches and the hydraulic control system, and because of the special tools and expertise required to perform an automatic transmission overhaul, it should not be undertaken by the home mechanic. Therefore, the procedures in this Chapter are limited to general diagnosis, routine maintenance, adjustment and transmission removal and installation.

If the transmission requires major repair work it should be left to a dealer service department or an automotive or transmission repair shop. You can, however, remove and install the transmission yourself and save the expense, even if the repair work is done by a transmission specialist.

2 Selector lever cable — adjustment

Refer to illustration 2.3

1   Place the gearshift selector lever in Park.
2   Working under the vehicle, push the transmission lever back against the stop to the Park position.
3   Loosen or remove the clip and adjust the cable so there is as little tension as possible at the ball socket **(see illustration)**.
4   Keep pressure on the shift lever and install the clip.
5   Check the gearshift selector lever in the Neutral and Drive positions to make sure it is within the confines of the lever stops. The engine must start only when the lever is in the Park or Neutral positions.

3 Selector lever cable — removal and installation

Refer to illustration 3.3

1   Disconnect the negative cable at the battery. Place the cable out of the way so it cannot accidentally come in contact with the negative terminal of the battery, as this would once again allow power into the electrical system of the vehicle.
2   Remove the exhaust system.
3   Disconnect the cable at the lock plate **(see illustration)**.
4   Bend back the metal clip holding the cable housing to the center

0026-H

2.3   There should be no tension on the transmission shift lever cable end (arrow) when the transmission is in Park

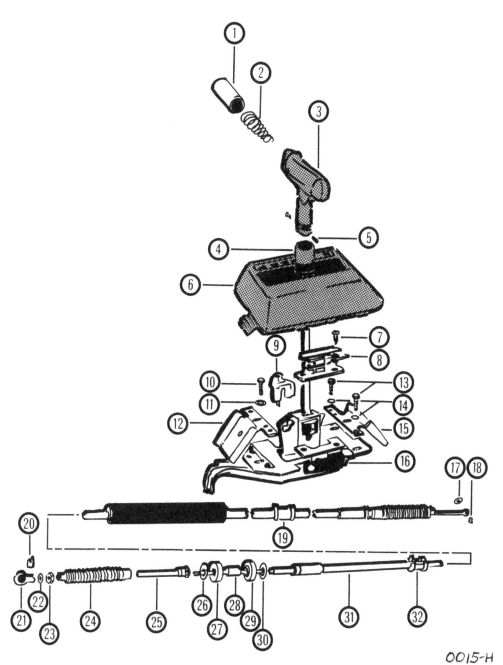

0015-H

3.3   Shift selector cable component layout

1 Push button	17 Rubber washer
2 Spring	18 Lock washer
3 Selector grip	19 Rear rubber grommet
4 Sleeve	20 Cable lockplate
5 Bolt	21 Ball socket
6 Cover	22 Washer
7 Screw	23 Nut
8 Contact plate	24 Dust cover
9 Bulb holder	25 Mounting tube
10 Bolt	26 Washer
11 Large washer	27 Rubber washer
12 Guide plate	28 Sleeve
13 Bolt	29 Rubber washer
14 Small washer	30 Washer
15 Guide plate	31 Selector lever cable
16 Shift mechanism	32 Front rubber grommet

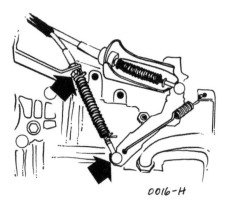

0016-H

4.2   Adjust the throttle pressure cable at the bracket and transmission lever (arrows) so the cable is free of tension

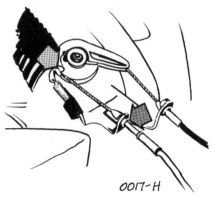

0017-H

4.4   Adjust the cable nut (right arrow) until the cable end is in the throttle cam (left arrow) with no tension on it

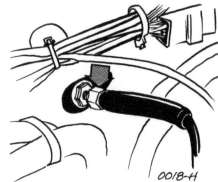

0018-H

4.5   The throttle pressure cable is adjusted at the nut on the firewall (arrow)

0019-H

4.6   To check for proper full throttle adjustment, push the pedal open until the first resistance is felt; the dog (arrow) should not be lifted off throttle cam

0020-H

4.7   For the kickdown test, push the throttle open and make sure the dog (arrow) is fully lifted off the throttle cam

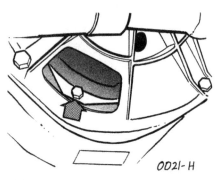

0021-H

7.8   The torque converter bolts (arrow) are accessible through the inspection hole

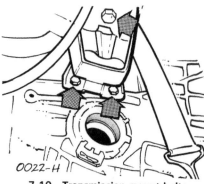

0022-H

7.10   Transmission mount bolts (arrows)

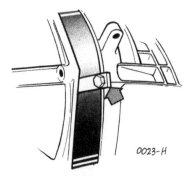

0023-H

7.11   The torque converter can be held in place during removal with short pieces of metal (arrow) bolted to the transmission

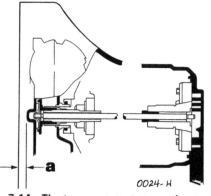

0024-H

7.14   The torque converter must have the specified clearance (a) before installation of the transmission

tunnel.

5   Remove the selector lever knob, squeeze the cover and pull it out of the center console.

6   Unplug and remove the bulb holder assembly.

7   Mark the location of the shift mechanism so that it can be reinstalled to the same position.

8   Unbolt the shift mechanism retaining bolts and guide plates.

9   Lift the front of the mechanism and pull the entire assembly forward. Disconnect the selector cables at the base and at the bottom of the selector lever.

10   Attach a long piece of wire or strong string to the forward end of the selector cable and pull the cable out of the vehicle from the rear. Leave the wire or string in place so you can pull the new cable into position.

11   Place the rubber grommet on the cable sleeve and coat the front grommet with rubber lubricant.

12   Install the lock plate on the cable.

13   Attach the new cable to the wire or string and pull it through into the passenger compartment while an assistant works from the rear to feed it through the insulation sheet. Pull the cable forward sufficiently to allow room to install the cable on the shift mechanism.

14   Install the cable end at the bottom of the selector lever and hook the other end to the bellows in the mount on the selector mechanism. Install the rubber washer between the cable end the selector lever.

15   Install selector mechanism and guide plates.

16   Install the bulb holder assembly.

17   Connect the lock plate to the cable.

18   Adjust the rubber grommets on the cable covers so the cable does not contact the central tube, then close the clamp.

19   Adjust the selector cable (Section 2).

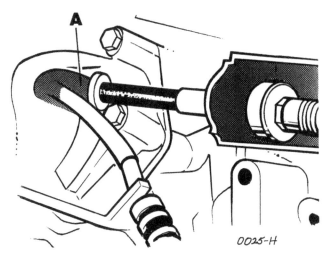

7.22 The selector lever cable grommet (A) must be installed securely to prevent damage to the cable

4 Throttle pressure cable — check and adjustment

Refer to illustrations 4.2, 4.4, 4.5, 4.6 and 4.7

1   The throttle pressure cable controls valves in the transmission which govern shift quality and speed. If shifting is harsh or erratic, the cables must be adjusted.
2   At the transmission, loosen the cable sleeve mounting nut on the bracket and the ball socket nut at the lever so they are free of tension but not slack **(see illustration)**.
3   Adjust the short cable sleeve on the firewall and the long cable sleeve on the bracket for the throttle cam plate until they are free of tension.
4   Adjust the cable sleeve nut until the cable end is positioned in the throttle cam slot with no tension on it **(see illustration)**.
5   Adjust the throttle pedal cable control at the firewall so this cable is also free of tension but not loose **(see illustration)**.
6   Check for full throttle adjustment by having an assistant push down on the throttle pedal until the first noticeable resistance is felt. The throttle should be wide open at this point and the dog in the throttle cam slot will not be lifted off **(see illustration)**.
7   To check the kickdown adjustment, have an assistant press the throttle pedal down fully and make sure the dog in the cam plate has lifted off **(see illustration)**. The operating lever should also be at or within 1° of full stop.

5 Throttle pressure cable — removal and installation

1   Disconnect the negative cable at the battery. Place the cable out of the way so it cannot accidentally come in contact with the negative terminal of the battery, as this would once again allow power into the electrical system of the vehicle.
2   Raise the vehicle and support it securely on jackstands.
3   In the passenger compartment, detach the throttle pressure cable from the throttle pedal and use a screwdriver to push the cable guide out of the firewall into the engine compartment.
4   In the engine compartment, disconnect the pressure cable at the throttle bracket.
5   Remove the exhaust system and the front and center crossmembers.
6   Under the rear of the vehicle, disconnect the throttle pressure cable at the transmission lever by unscrewing the ball socket and removing the lock pin.
7   If the cable is to be reinstalled, mark the positions where the hose clamps and cable sleeve retain the cable to the drive tube so it can be reinstalled to the same position. Remove the hose clamps and disconnect the cable from the sleeve.
8   Remove the cable by pulling it downward between the central tube

and the insulation sheet with a wire hook, using a putty knife or similar tool to carefully push back the sheet.
9   Prior to installation, coat the cable with rubber lubricant where it contacts the insulation sheet.
10   Place the ends of the cable in position and carefully press it into place in the insulation sheet, using the putty knife.
11   Connect the cable ends.
12   The remainder of installation is the reverse of removal.
13   After installation, adjust the cable (Section 4).

6 Differential axleshaft flange seal — replacement

Refer to Section 3, Part A. The procedure is the same as for manual transmissions.

7 Automatic transmission — removal and installation

Refer to illustrations 7.8, 7.10, 7.11, 7.14 and 7.22

Removal

1   Disconnect the negative cable at the battery. Place the cable o t of the way so that it cannot accidently come into contact with tl ; terminal, which would again allow current flow.
2   Raise the vehicle and support it securely on jackstands.
3   Remove the exhaust system.
4   Detach the driveaxles from the transmission and suspend them out of the way horizontally on pieces of wire so the CV joint boots are not damaged.
5   Disconnect and plug the fluid cooler tubes at the transmission.
6   On later models, disconnect the speedometer cable from the transmission.
7   Disconnect selector lever and transmission lever cables. The cable has a locking ball at its head to retain the cable.
8   Remove the inspection hole cover from the bellhousing for access to the torque converter **(see illustration)**. Mark the torque converter-to-driveplate relationship so they can be reinstalled in the same position. Remove the three torque converter mounting bolts in turn, using a wrench on the front pulley to turn the engine.
9   Support the transmission with a jack.
10   Raise the transmission slightly and remove the retaining bolts from the mounts and bellhousing **(see illustration)**.
11   The torque converter must remain with the transmission during removal. Retainers can be made from short pieces (approximately 4 inches long) of strap metal with holes drilled in the ends so they can be bolted to the transmission **(see illustration)**. Alternately, an assistant can hold the torque converter in position during removal.
12   Carefully move the transmission to the rear and lower it from the vehicle.
13   Lower the transmission onto a clean working surface and remove the torque converter.

Installation

14   Prior to installation, slide the torque converter onto the transmission input shaft splines until it can be felt touching bottom. Push in on the one-way clutch carefully, taking care not to tilt it while moving it back and forth slightly to mesh the splines. When fully inserted against the stop, measure the distance between the end of the torque converter and the edge of the bellhousing to make sure there is the specified clearance **(see illustration)**.
15   Raise the transmission carefully into position and align the torque converter and driveplate bolt holes.
16   Install the transmission mount-to-body bolts finger tight.
17   Install the transmission bellhousing bolts finger tight.
18   Tighten the transmission mount bolts to the specified torque.
19   Tighten the transmission bellhousing bolts to the specified torque.
20   Install the torque converter-to-driveplate bolts finger tight.
21   Tighten the torque converter bolts in turn to the specified torque.
22   Connect selector lever and transmission lever cables, making sure the rubber grommet is securely mounted between the selector cable cable and the suspension mount **(see illustration)**.
23   Connect the transmission cooler fluid lines.
24   On later models, connect the speedometer cable.
25   Connect the driveaxles.

Chapter 8 Clutch and drivetrain

Contents

Specifications

Clutch slave cylinder release lever travel	0.59 in (15 mm) to 0.71 in (18 mm)
minimum allowable travel	0.59 in (15 mm)
Clutch pedal free play	0.12 in (3 mm)
Clutch preload spring plate-to-bearing distance	
(A in illustration 2.3)	2.41 in (61 mm)

Torque specifications

	Ft-lbs	Nm
Central tube-to-transmission bolt		
12 mm	61	83
10 mm	30	41
Central tube-to-bellhousing bolt	30	41
Clutch pressure plate bolt	18	24
Clutch master cylinder nut	15	20
Bellhousing-to-engine bolt	54	73
Driveaxle bolt	30	41
Driveshaft-to-transmission input shaft clamp bolt	58	79
Engine speed sensor bolt	6	8
Flywheel-to-crankshaft bolt	66	89
Slave cylinder bolt	15	20

1 General information

The information in this Chapter deals with the components from the rear of the engine to the rear wheels, except for the transmission (see Chapter 7). For the purposes of this Chapter, these components are grouped into three categories; clutch, driveline including the central tube and driveshaft assembly, and the driveaxles.

The clutch disc is held in place against the flywheel by the pressure plate springs. During disengagement, such as during gear shifting, the clutch pedal is depressed, which operates the hydraulic slave cylinder, pushing the release bearing on the pressure plate springs, disengaging the clutch.

The rear mounted transmission is connected to the bellhousing, which is bolted to the engine, by the central tube and driveshaft as-

sembly. The driveshaft, located inside the central tube, transfers the power from the clutch to the transmission.

Power is passed from the transmission to the wheels by the driveaxles. The driveaxles consist of solid shafts with constant velocity (CV) joints at each end. The CV joints are internally splined and contain ball bearings which allow them to transfer power smoothly at various lengths and angles as the driveaxles move through their full range of travel. The CV joints are lubricated with special grease and are protected by rubber boots which must be inspected periodically for cracks, tears and signs of leakage, which could lead to damage of the joints and failure of the driveaxle.

As nearly all the procedures covered in this Chapter involve working under the vehicle, make sure it is firmly seated on sturdy jackstands or on a hoist where the vehicle can be easily raised and lowered.

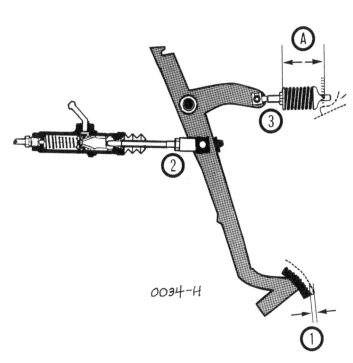

0034-H

2.3   The clutch pedal free play is measured at the pedal pad (1) and adjusted at the pushrod nut (2). The preload spring distance (A) is adjusted at the nut (3)

3.4a   Measure the movement at the end of the slave cylinder piston, first in the released . . .

3.4b   . . . and then in the fully extended position to determine the release travel distance

2   Clutch free play and preload spring — check and adjustment

Refer to illustration 2.3

Free play

1   The correct clutch free play is very important for proper clutch operation and to ensure normal clutch service life.
2   To check the free play, slowly press the clutch pedal until resistance by the clutch hydraulic cylinder is felt.
3   Measure the distance the clutch pedal has traveled and compare it to Specifications. If adjustment is required, loosen the locknut on the clutch pedal pushrod and turn the rod until the specified free play is obtained. Tighten when the adjustment is completed **(see illustration)**.

Preload spring

4   A preload spring is used on these models to reduce the amount of effort required when depressing the clutch pedal.
5   To check the adjustment of the preload spring, measure the distance between the outer edge of the plate and bearing with the clutch pedal in the released position **(see illustration 2.3)**.
6   Adjust the preload by turning the nut on the front of the spring until the specified measurement is achieved.

3   Clutch release travel — check and adjustment

Refer to illustrations 3.4a and 3.4b

1   The clutch slave cylinder release lever must have sufficient travel to fully disengage the clutch and this travel can be measured and checked against Specifications.
2   Raise the vehicle and support it securely on jackstands.
3   Remove the clutch plate inspection plug from the bellhousing.
4   Measure the distance the end of the slave cylinder piston moves between the released and fully extended position **(see illustrations)**. Compare this measurement to the Specifications.
5   If the release travel is less than specified, the clutch is not fully

disengaging. Possible causes of insufficient travel are air in the hydraulic system, faults in the clutch master and/or slave cylinders, a worn clutch plate or carpeting bunched up under the pedal.

4   Clutch components — removal and installation

Refer to illustrations 4.8, 4.10, 4.14, 4.19, 4.25a, 4.25b, 4.25c, 4.27a, 4.27b, 4.28, 4.30a, 4.30b, 4.31, 4.32, 4.33, 4.34, 4.38 and 4.40

Removal

1   Access to the clutch components is normally accomplished by removing the transmission, disconnecting the central tube and moving it back sufficiently for access to the clutch, leaving the engine in the vehicle. If the engine is being removed for major overhaul, then the opportunity should always be taken to check the clutch for wear and replace worn components as necessary. The following procedures will assume that the engine will stay in place.
2   Disconnect the negative cable at the battery. Place the cable out of the way so it cannot accidentally come in contact with the negative terminal of the battery, as this would once again allow power into the

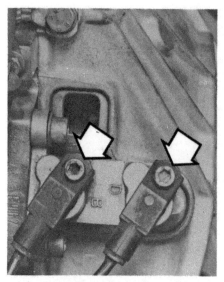

4.8   The engine speed sensors (arrows) are located at the top left of the clutch housing

4.10   The wiring harness clamp nut is located just below the engine speed sensors

4.14   Move the clutch slave cylinder out of the way. Be careful not to damage the fluid line

electrical system of the vehicle.

3   Remove the exhaust system.

4   Raise the vehicle and support it securely on jackstands.

5   Referring to Chapter 7 Part A, remove the transmission from the vehicle.

6   Support the rear end of the central tube with a jack or jackstand so that it does not rest on the torsion bar housing.

7   Remove the starter.

8   Remove the Allen head bolts retaining the engine speed sensors located on the bellhousing at the rear of the engine, adjacent to the firewall **(see illustration)**. Remove the sensors and move them out of the way.

9   Disconnect the ground cable at the firewall.

10   Remove the retaining nut and disconnect the wiring harness clamp from the bellhousing **(see illustration)**.

11   In the engine compartment, remove the two upper bellhousing bolts.

12   From under the vehicle, remove the lower two central tube-to-bell-housing bolts.

13   Support the engine with a jack under the oil pan, using a block of wood to protect the pan.

14   Remove the clutch slave cylinder and move it out of the way without disconnecting the fluid line **(see illustration)**.

15   Remove the upper two bellhousing-to-engine bolts.

16   Remove the upper central tube retaining bolts.

17   Remove the rear support and pry the central tube rearward to allow sufficient clearance for bellhousing removal.

18   Remove the bellhousing guard.

19   Remove the clutch release lever shaft retaining bolt and screw an 8-mm bolt into the end of the shaft **(see illustration)**. Pull the shaft out by prying on the bolt with a screwdriver.

20   Remove the lower two retaining bolts and lower the bellhousing from the engine.

21   An internal tooth socket will be necessary when removing the pressure plate-to-flywheel bolts. Turning each bolt only one turn at a time, slowly loosen the bolts. Work in a diagonal pattern, loosening only a little at a time until all spring pressure is relieved. Remove the pressure plate and clutch disc.

Inspection

22   Ordinarily, when a fault is found in the clutch system, it can be attributed to wear of the clutch driven plate assembly (clutch disc). However, all components should be inspected at this time.

23   Inspect the flywheel as discussed in the next Section.

24   Inspect the pilot bearing (see Section 6).

25   Inspect the facing on the clutch disc for wear and damage **(see illustration)**. There should be at least 1/16-inch of lining above the rivet heads **(see illustration)**. Check for loose rivets, distortion, cracks,

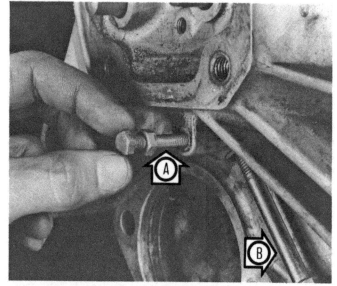

4.19   Remove the release lever retaining bolt (A) and screw a bolt into the end of the release lever shaft (B)

broken springs or any other obvious damage. Inspect the rubber cushioning discs for cracking and separation **(see illustration)**. As mentioned above, ordinarily the disc is replaced as a matter of course, so if in doubt about the quality, replace with a new one.

26   Inspect the release lever and shaft for wear on the contact surfaces and looseness in the bearings.

27   Ordinarily, the release bearing is also replaced along with the clutch disc. Remove the snap-ring and lift off the release bearing **(see illustration)**. Turn the bearing and check for wear and roughness **(see illustration)**.

28   Check the machined surfaces of the pressure plate **(see illustration)**. If the surface is grooved or otherwise damaged, take it to a machine shop for possible machining or replacement. Also check for obvious damage, distortion, cracking, etc. If a new pressure plate is indicated, new or factory rebuilt units are available.

Installation

29   Before installation, carefully wipe clean the flywheel and pressure plate machined surfaces. It is important that no oil or grease is on these surfaces or the facing of the clutch disc. Handle these parts only with clean hands.

4.25a   Inspect the clutch disc lining for wear and damage. The surface here (arrows) is breaking up

4.25b   Measure the thickness of the clutch lining from the rivet head to the top surface of the lining. Here a tire depth gauge is used

4.25c   Check the rubber clutch cushion discs (arrows) for damage and separation

4.27a   Use snap-ring pliers to remove the release bearing from the pressure plate

4.27b   Turn the bearing to check for loosenss and rough rotation

4.28   Check the friction surface of the pressure plate for hot spots and scoring

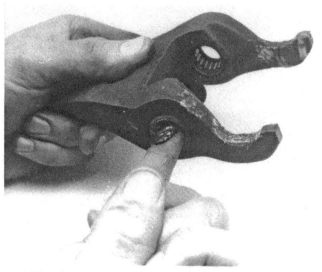

4.30a   Apply a small amount of grease to the release lever
bearings and . . .

4.30b   . . . the contact surfaces

4.31   Use your fingers to apply a small mount of grease to
the clutch splines

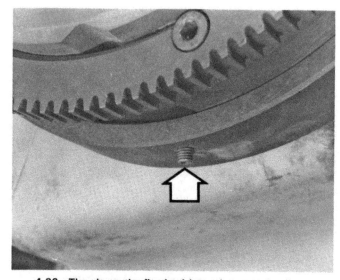

4.32   The pin on the flywheel (arrow) must point down
before the clutch housing is installed

4.33   Use an alignment tool when installing the clutch disc
and pressure plate

4.34   Install all the pressure plate bolts finger tight before
final tightening

4.38 With the release lever (arrow) in position, place the clutch housing over the dowels

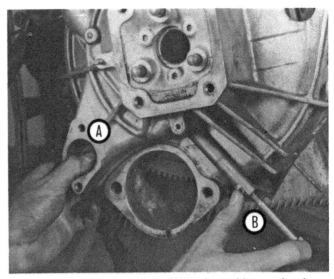

4.40 Hold the release lever (A) in place with your thumb and secure it with the shaft (B)

30 Lubricate the contact surfaces and bearings of the release lever with moly base grease (see illustrations).
31 Lubricate the splines of the clutch disc with a small amount of moly base grease (see illustration).
32 Make sure that the pin on the flywheel is pointing straight down, rotating the flywheel if necessary (see illustration).
33 Put the clutch disc and pressure plate in position with the clutch disc held in place with the alignment tool (see illustration).
34 Install the pressure plate-to-flywheel bolts and tighten them only finger tight, working around the pressure plate (see illustration).
35 If an alignment tool has not been used up to this point, center the disc by inserting a long screwdriver or metal bar through the middle of the clutch disc. Move the disc until it is exactly in the center and the splines of the driveshaft will pass easily through the disc and into the pilot bearing.
36 One turn at a time, tighten the pressure plate-to-flywheel bolts. Work in a diagonal pattern to prevent distorting the cover as the bolts are tightened to the proper torque specification.
37 Using moly base grease, lubricate the bellhousing guide sleeve.
38 Insert the release lever arm into the bellhousing, raise the housing into place and install it on the engine block dowels (see illustration).
39 Install the bolts and tighten them to the specified torque.
40 Hold the lever in position while guiding the shaft through it into place (see illustration).
41 Hold the shaft in position and install the lock bolt.
42 Lubricate the surface where the slave cylinder contacts the release arm with white lithium base grease.
43 Install the central tube, transmission, slave cylinder and all components removed previously, tightening all fasteners to the proper torque specifications.

5 Flywheel/driveplate — inspection, removal and installation

Refer to illustration 5.3

1   The flywheel as used on manual transmission vehicles, or the driveplate (flexplate) used on vehicles with an automatic transmission, is accessible only after removing the engine or the transmission and central tube.
2   Visually inspect the flywheel for cracks, heat checking or other obvious defects. If the imperfections are slight, a machine shop can machine the surface flat and smooth. Check the flywheel/driveplate starter teeth (ring gear) for wear and chipping.
3   To remove the flywheel/driveplate, remove the bolts which attach it to the engine using a suitable 12 point internal tooth socket (see illustration). Be aware that it is fairly heavy and should be well supported as the final bolt is removed — do not drop it.
4   Upon installation, note that the flywheel/driveplate fits over an

5.3 An internal tooth socket is necessary to remove the flywheel bolts

alignment dowel so that it will go on only one way. Tighten the attaching bolts to the proper torque specification.

6 Pilot bearing — removal, inspection and installation

Refer to illustrations 6.7a, 6.7b and 6.8

1   The clutch pilot bearing is a needle-type bearing which is pressed into the rear of the crankshaft. Its purpose is to support the front of the driveshaft. The pilot bearing should be inspected whenever the clutch components are removed from the engine. Due to its inaccessibility, if you are in doubt as to its condition, replace it with a new one. Note: *If the engine has been removed from the vehicle, disregard the following steps which do not apply.*
2   Remove the clutch components (Section 4).
3   Using a clean rag, wipe the bearing clean and inspect for any excessive wear, scoring or obvious damage. A flashlight will be helpful to direct light into the recess.
4   Removal can be accomplished with a special hooked puller tool, but an alternative method also works very well.

6.7a   Insert the bolt into the bearing, and hook the head behind it

6.7b   Tap the locking pliers with a hammer to dislodge the bearing from the bore

6.8   Use a socket and extension and a hammer to tap the pilot bearing evenly into the crankshaft bore

7.4a   Unscrew the slave cylinder fluid line . . .

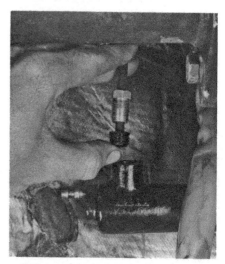

7.4b   . . . and press the bleed screw cover over the end of the line to plug it

5   Find a long bolt, the head of which is slightly smaller in diameter than the opening in the bearing, and fasten locking pliers to the threaded end.

6   Check the bolt head for fit — it should just slip into the bearing with very little clearance.

7   Insert the head of the bolt into the bore, hook it on the back of the bearing securely and lightly tap with a hammer on the locking pliers to dislodge the bearing (see illustrations). Remove the bearing and clean out the bore in the crankshaft recess.

8   To install the new bearing, lubricate the outside surface with oil then drive it evenly into the recess with a suitable size socket and extension and a hammer (see illustration).

9   Install the clutch components, transmission and all other components removed previously, tightening all fasteners properly.

---

7   Clutch slave cylinder — removal, overhaul and installation

---

Refer to illustrations 7.4a, 7.4b, 7.5, 7.7, 7.8a, 7.8b, 7.8c and 7.10

Note: *Before beginning this procedure, contact local parts stores or dealers concerning the purchase of a rebuild kit or a new slave cylinder. Availability and cost of the necessary parts may dictate whether the cylinder is rebuilt or replaced with a new one. If it is decided to rebuild the cylinder, inspect the inside bore as described in Step 9 before purchasing parts.*

Removal

1   Disconnect the negative cable at the battery. Place the cable out of the way so that it cannot accidently come into contact with the terminal which would again allow current to flow.

2   Raise the vehicle and support it securely on jackstands.

3   Remove the starter motor.

4   Disconnect the fluid line at the slave cylinder (see illustration). Have a small can and rags handy, as some fluid will be spilled as the line is removed. Plug the line, using the rubber bleed screw cover (see illustration).

5   Remove the two mounting bolts (see illustration).

6   Remove the slave cylinder.

Overhaul

7   Pry off the retaining circlip with a small screwdriver (see illustration).

8   Withdraw the pushrod assembly, piston and spring from inside the cylinder (see illustrations).

9   Carefully inspect the inside bore of the cylinder. Inspect for deep scratches, scores or ridges. The bore must be smooth to the touch. If any imperfections are found, the slave cylinder must be replaced with a new one.

10   Using the new parts in the rebuild kit, assemble the components, using plenty of fresh brake fluid for lubrication. Note the proper direction of the spring and the seal. Use a suitable socket to seat the circlip by pressing it in evenly (see illustration).

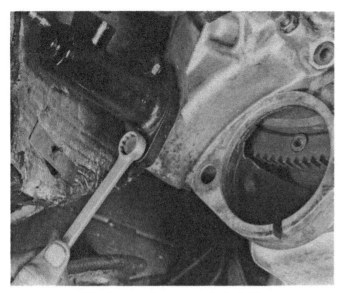

7.5   Remove the slave cylinder bolts

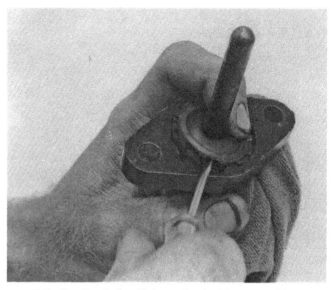

7.7   Pry the circlip off the pushrod with a screwdriver

7.8a   Remove the pushrod . . .

7.8b   . . . piston and . . .

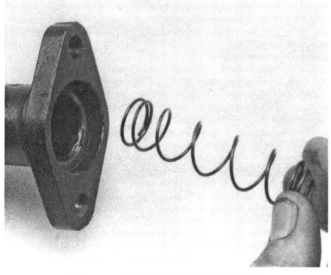

7.8c   . . . spring from the cylinder bore

7.10   Seat the circlip by pushing it in evenly with a socket

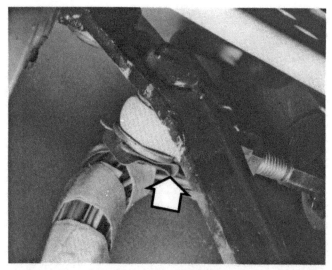

8.1   The pushrod is retained on the clutch pedal arm with
a clip (arrow)

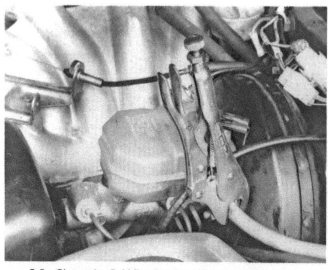

8.3   Clamp the fluid line between the reservoir and the
clutch master cylinder with locking pliers

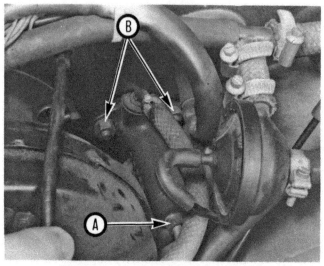

8.4   Unscrew the fluid pressure line at the front of the
master cylinder (A) and remove the retaining bolts (B)

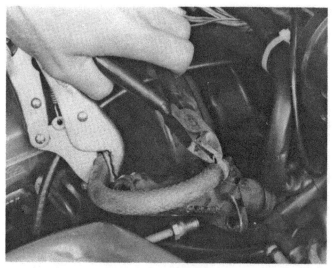

8.6a   Cut the fluid feed hose clamp with wire cutters . . .

8.6b   . . . and pull the hose off

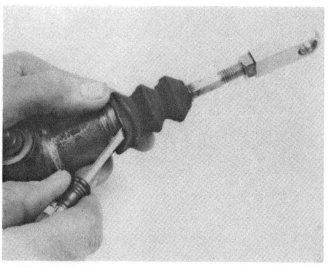

8.8   Insert a small screwdriver under the boot and work
around the circumference to dislodge it from the
master cylinder

8.10a   Hold the piston in while removing the snap-ring

8.10b   Withdraw the piston from the bore

Installation

11 Lubricate the clutch pushrod contact surfaces with white lithium base grease.
12 Install the slave cylinder on the bellhousing, tightening the bolts to the proper torque.
13 Connect the hydraulic line to the slave cylinder.
14 Fill the clutch master cylinder with brake fluid conforming to DOT 3 specifications.
15 Bleed the system as described in Section 9.
16 Lower the vehicle and connect the negative battery cable.

8   Clutch master cylinder — removal, overhaul and installation

Refer to illustrations 8.1, 8.3, 8.4, 8.6a, 8.6b, 8.8, 8.10a and 8.10b

1 In the passenger compartment, remove the retaining clip and disconnect the clutch master cylinder pushrod from the clutch pedal **(see illustration)**.
2 In the engine compartment, move any vacuum hoses which would interfere with removal of the master cylinder.
3 Use locking pliers to clamp the master cylinder fluid feed line **(see illustration)**.
4 Disconnect the hydraulic fluid pressure line at the slave cylinder and remove the two retaining nuts **(see illustration)**.
5 Pull the master cylinder away from the firewall.
6 Cut the fluid hose clamp and pull the hose off the master cylinder **(see illustrations)**.
7 Remove the master cylinder from the vehicle.
8 Pry the boot out of the groove in the master cylinder with a small screwdriver and remove the pushrod **(see illustration)**.
9 Place the master cylinder securely in a vise, using blocks of wood or thick cloth to protect the surface.
10 Push the piston in with a punch or similar tool and remove the snap-ring. Withdraw the piston **(see illustrations)**.
11 Carefully inspect the inside bore of the cylinder. Check for deep scratches, scores or ridges. The bore must be smooth to the touch. If any imperfections are found, the master cylinder must be replaced with a new one.
12 Using the new parts in the rebuild kit, assemble the components using plenty of fresh brake fluid for lubrication. Note the proper direction of the spring and the seal and make sure the snap-ring is securely installed in the groove.
13 The remainder of installation is the reverse of removal.
14 After installation, bleed the clutch hydraulic system (Section 9).

9   Clutch hydraulic system — bleeding

Refer to illustration 9.6

1 The hydraulic system should be bled of all air whenever any part of the system has been removed or if the fluid level has been allowed to fall so low that air has been drawn into the master cylinder. The procedure is very similar to bleeding a brake system because on these models the clutch and brake systems use the same fluid reservoir.
2 Fill the master cylinder with new brake fluid conforming to DOT 3 specifications. **Caution:** *Do not re-use any of the fluid coming from the system during the bleeding operation or use fluid which has been inside an open container for an extended period of time.*
3 Raise the vehicle and place it securely on jackstands to gain access to the slave cylinder, which is located on the left side of the bellhousing.
4 Remove the starter motor.
5 Remove the dust cap which fits over the bleeder valve and push a length of plastic hose over the valve. Place the other end of the hose in a clear jar with about two inches of brake fluid. The hose end must be in the fluid at the bottom of the jar.
6 Have an assistant depress the clutch pedal and hold it. Open the bleeder valve on the slave cylinder, allowing fluid to flow through the hose **(see illustration)**. Close the bleeder valve when your assistant signals that the clutch pedal is at the bottom of its travel. Once closed, have your assistant release the pedal.

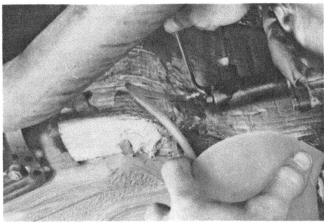

9.6   Open the bleed screw while an assistant holds the clutch pedal down to bleed the system

10.2   An internally splined socket extension is necessary for removing the driveaxle bolts

10.3   Support the CV joints when lowering the driveaxle

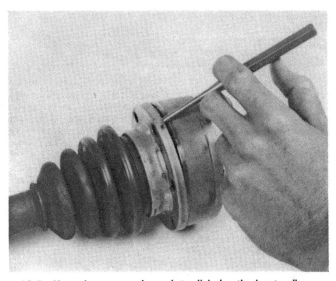

10.5   Use a hammer and punch to dislodge the boot collar from the CV joint

10.6   Mark the CV joint component relationship before disassemby

7   Continue this process until all air is evacuated from the system, indicated by a full, solid stream of fluid being ejected from the bleeder valve each time and no air bubbles in the hose or jar. Keep a close watch on the fluid level inside the master cylinder. If the level drops too low, air will be drawn into the system and the process will have to be started all over again.

8   Install the starter motor and lower the vehicle. Check carefully for proper operation before placing the vehicle in normal service.

---

10   Driveaxle, constant velocity (CV) joints and boots — removal, disassembly, inspection and installation

---

Refer to illustrations 10.2, 10.3, 10.5, 10.6, 10.7, 10.8, 10.9a, 10.9b, 10.10, 10.15, 10.18, 10.19 and 10.20

1   Raise the vehicle and support it securely on jackstands.
2   Using a suitable Allen head socket, remove the driveaxle bolts **(see illustration)**.
3   Grasp the driveaxle securely, supporting the CV joints **(see illustration)**, and lower it from the vehicle, noting the direction in which it is installed.
4   Place the driveaxle on a clean working surface.

5   Use a punch and hammer to work around the outer circumference and dislodge the collar from the joint housing **(see illustration)**.
6   Paint or scribe across the components so they can be reinstalled in the same relationship **(see illustration)**.
7   Remove the circlip from the end of the axle **(see illustration)**.
8   Push the axleshaft out of the joint with your thumbs **(see illustration)**.
9   To remove the boot, note the direction in which it is installed, remove the concave washer from the axleshaft with pliers and then slide the boot off the end of the axleshaft **(see illustrations)**.
10   Align the ball bearings with the grooves in the outer race and separate the outer race from the inner **(see illustration)**.
11   Press the roller balls out of the race.
12   Wash the components in solvent and dry them thoroughly.
13   Inspect the roller ball bearings, splines and races for damage, corrosion, wear and cracks.
14   If any of the components are not serviceable, the entire CV joint must be replaced with a new one.
15   Press the ball bearings into the race until they snap into place **(see illustration)**.
16   Reassemble the inner and outer bearing races.
17   Install the boot on the shaft, carefully guiding it over the splines.
18   Slide the concave washer onto the shaft splines and press it into position with a suitable deep socket **(see illustration)**.

10.7   Use circlip pliers to expand the circlip and lift it out of the groove

10.8   Hold CV joint securely and push the axleshaft out with your thumbs

10.9a   Remove the concave washer by working it up the shaft with pliers

10.9b   Slide the boot off

10.10   Tilt the inner race out of the outer race and separate them

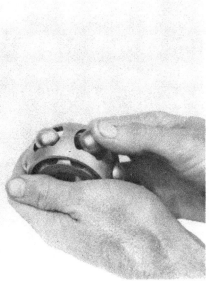

10.15   Press the bearings in until they snap into place

10.18   Use a hammer and deep socket to tap on the concave washer until it is at the base of the splines

10.19   Press the grease into the boot with your fingers

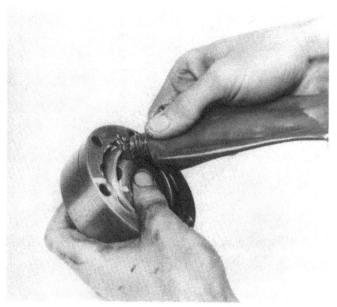

10.20   Force the grease into the rollers and races

19 Place a small amount of moly base CV joint grease into the boot (see illustration).
20 Pack the bearing assembly with moly base CV joint grease (see illustration).
21 Lubricate the splines with a small mount of grease and slide the CV joint into place, making sure the marks made during removal are aligned.
22 Secure the joint with the circlip and tap the boot collar onto the joint groove.
23 Place the driveaxle in position and install the bolts finger tight. Tighten the bolts to the specified torque.
24 Lower the vehicle.

Chapter 9 Brakes

Contents

Specifications

Brake lining thickness wear limit	0.080 in (2 mm)
Brake rotor thickness wear limit	0.800 in (20.0 mm)
Brake rotor lateral runout	0.002 in (0.05 mm) maximum
Brake rotor lateral runout installed	0.004 in (0.10 mm) maximum
Brake pedal minimum play (engine off)	0.400 in (10.16 mm)
Brake pedal-to-brake light switch clearance	0.250 in (5 mm)
Brake pedal pushrod adjustment length	7.250 in (186 mm)
Parking brake drum diameter wear limit	7.125 in (181 mm)

Torque specifications

	Ft-lbs	Nm
Spindle clamping nut Allen bolt	10 to 12	13 to 16
Caliper-to-steering knuckle bolts	63	85
Brake disc-to-hub nut	17	23
Rear wheel hub-to-shaft		
with steel trailing arm	280 to 332	380 to 450
with aluminum trailing arm	340	460
Brake line holder screw	7.5	10
Brake disc-to-hub screw	3.7	5
Caliper-to-brake backing plate or trailing arm bolt	63	85
Brake backing plate-to-trailing arm bolt	43	58
Brake cable-to-yoke	6.3	8.5
Brake booster-to-adapter	16	21
Brake booster adapter-to-firewall	16	21
Swivel on brake pushrod	26	35

1 General information

General

The brake system used on the Porsche 944 is a four wheel disc, power assisted system. The emergency brake is a mechanically operated drum type brake fitted to the rear wheels.

All models covered in this manual are equipped with single or four-piston calipers and ventilated rotors. The hydraulic system is a split design, meaning there are separate circuits for the front and back wheels. If one circuit fails, the other circuit will remain functional and allow straight line braking until the vehicle can be repaired. Never drive a vehicle with a malfunctioning brake circuit.

Warning: *Later models equipped with ABS brakes use various components specifically designed for this system. Repair work on ABS systems is not recommended for the home mechanic.*

Master cylinder

The master cylinder is located on the engine firewall, under the hood, and is easily recognized by the fluid reservoir on top. The hydraulic clutch uses fluid from the same reservoir and should, therefore, always be bled at the same time as the brakes. Porsche uses two different master cylinders, manufactured by either Girling or Teves. The Teves version is stamped with the symbol Ate and the Girling version with a G. This identification is not present on all master cylinders, though, so it's best to take the unit to the parts store for comparison. When ordering parts for the master cylinder, be sure you know which one you've got.

Since power brakes are standard, a power booster is located adjacent to the master cylinder. Connected to the brake pedal is a pushrod, which runs through the power booster and operates the components inside the master cylinder.

Disc brake assemblies

Each of the wheels has a single or four-piston caliper. Hydraulic pressure created by pushing on the brake pedal forces the piston(s) to move, in effect clamping against the moving rotor.

Replaceable pads – two per wheel – are the components which actually press against the rotor. They will, over a period of time, wear out. The 1984 models have an electronic wear sensor to tell you when the pads need to be replaced. Other models, including the later ones, have a simpler mechanical system which makes a screeching noise when the pads are worn out.

Parking brake

Located inside the rear rotor assemblies, the parking brake resembles a conventional drum brake. When the parking brake lever is applied, the linings are forced against the inside of the rotor housing, locking the rear wheels.

Unless the parking brake system has been abused (as in driving with the brake partially engaged), the system components do not need frequent periodic inspection and maintenance. If, however, the vehicle does not hold firmly on a hill with the parking brake applied, the system components should be carefully inspected.

Precautions

There are some general notes and cautions involving the brake system on this vehicle:

a) Use only brake fluid specified for this system. See *Recommended fluids and lubricants* in Chapter 1 for the correct fluid.

b) The brake pads and linings contain asbestos fibers which are hazardous to your health if inhaled. Whenever you work on brake system components, carefully wipe all parts clean with a water or alcohol dampened cloth. Do not allow the fine dust to become airborne.

c) Safety should be paramount whenever any servicing of the brake components is performed. Do not use parts or fasteners which are not in perfect condition, and be sure that all clearances and torque specifications are adhered to. If you are unsure about a certain procedure, seek professional advice. Upon completion of any brake system work, test the brakes carefully in a controlled area before driving the vehicle. If a problem is suspected in the brake system, do not drive the vehicle until the fault is corrected.

2   Brake pad wear sensor – inspection and replacement

1   1984 models incorporate a small, cylindrical plastic wear sensor in the caliper assembly which houses and insulates the wire going to the wear indicator light on the dashboard. If the pads wear to a thickness of 2.0 mm, the rotor will shave off the insulation and make contact with the wire, causing the light to flash. Whenever it does this, the wear sensor, as well as the pads, must be replaced.

2   To replace the sensor, remove the wheel and look at the backside of the caliper. Disconnect the wire electrical connector, remove the wear sensor and replace it with a new one. Be sure you've got the sensor pushed in all the way. This operation can most easily be performed as part of the brake pad replacement in Section 3 of this Chapter.

3   Brake pads – removal and installation

Refer to illustrations 3.4a through 3.4g

Note: *The following information applies to both the front and rear brake assemblies. Some later high-performance models use four-piston brake calipers front and rear. Due to the complexity of this design and the special tools required for servicing, we recommend taking the vehicle to a dealer service department for this procedure.*

Warning: *Disc brake pads must be replaced on both wheels at the same time – never replace the pads on only one wheel. Also, brake system dust contains asbestos, which is extremely harmful to your health. Never blow it out with compressed air and do not inhale any of it. Do not, under any circumstances, use gasoline or petroleum-based solvents to clean brake parts. Use brake cleaner or denatured alcohol only.*

1   Remove about two-thirds of the fluid from the master cylinder reservoir(s) with a syringe.

2   Raise the vehicle and support on jackstands. Remove the wheels.

3   Inspect the rotor carefully as outlined in Section 5. If machining is necessary, follow the information in that Section to remove the rotor, at which time the pads can be removed from the caliper as well.

4   Follow the accompanying photos, beginning with illustration 3.4a. Stay in order and read the caption under each illustration.

5   Installation is the reverse of removal. Replace brake pads which have deep cracks, have become loose from the backing plates or are

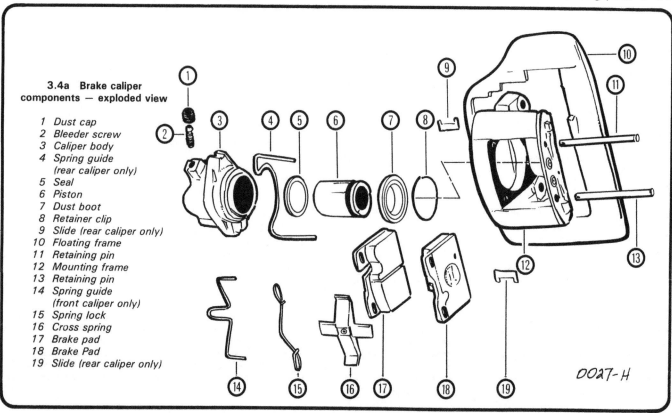

3.4a   Brake caliper components — exploded view

1   Dust cap
2   Bleeder screw
3   Caliper body
4   Spring guide (rear caliper only)
5   Seal
6   Piston
7   Dust boot
8   Retainer clip
9   Slide (rear caliper only)
10 Floating frame
11 Retaining pin
12 Mounting frame
13 Retaining pin
14 Spring guide (front caliper only)
15 Spring lock
16 Cross spring
17 Brake pad
18 Brake Pad
19 Slide (rear caliper only)

0027-H

covered with oil. When replacing the retainer spring, be sure the ends are completely seated in the retainer pins.

6   Once the new pads are in place, install the wheels and lower the vehicle to the ground. **Note:** *If the fluid inlet fitting was disconnected from the caliper for any reason, the brake system and the clutch must*

be bled as described in Section 15.

7   Fill the master cylinder reservoir(s) with new brake fluid and slowly pump the brakes a few times to seat the pads against the rotor.

8   Check the fluid level in the master cylinder reservoir(s) one more time and then road test the vehicle carefully before driving it in traffic.

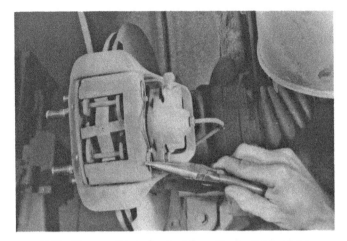

3.4b   Remove the spring lock from the backside of the caliper. This will allow the pad retaining pins to be removed

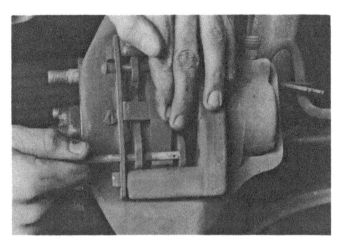

3.4c   Remove the pad retaining pins. If they're stuck, try driving them out with a drift punch and hammer

3.4d   Compress the piston into the cylinder as far as it will go. Do not apply pressure directly to the piston and do not use the brake disc for leverage

3.4e   Pull the inner pad out for inspection or replacement. Note the wear indicator slot here: this pad is still usable

3.4f   The outer pad will still be pressed between the caliper frame and the disc. Make more room by prying with a screwdriver

3.4g   The outer pad can now be easily removed

4   Caliper – removal, overhaul and installation

Refer to illustrations 4.3a, 4.3b, 4.4, 4.7, 4.8, 4.9, 4.10, 4.11, 4.14a, 4.14b, 4.15, 4.17, 4.19a, 4.19b, 4.19c, 4.20, 4.21, 4.24 and 4.26

Note: *The following procedure applies to most calipers used on these models. Where different designs are encountered (as in the later four-piston caliper design), much of the following is valid, however, use the instructions which accompany the rebuild kit for any differences. New and factory rebuilt calipers are available on an exchange basis, which makes this job quite easy. If it is decided to rebuild the calipers, make sure that a rebuild kit is available before proceeding. Always rebuild the calipers on all wheels - never rebuild just one of them.*

Removal

1   Raise the vehicle and support it securely on jackstands. Remove the wheel.
2   Remove the brake pads and set them aside (Section 3).
3   Disconnect the hose retainer **(see illustration)** and loosen the two bolts which attach the caliper to the steering knuckle **(see illustration)** or, on the rear brakes, the brake backing plate.
4   Loosen but do not remove the hydraulic line nut at the caliper body **(see illustration)**.
5   Remove the two bolts which attach the caliper to the steering knuckle (front) or brake backing plate (rear).

6   With plenty of rags handy to capture escaping fluid, disconnect the hydraulic line. Be careful not to twist or bend it, as this may cause internal damage to the rubber and eventual brake failure. If in doubt about the condition of the brake lines, replace them with new ones.
7   Plug the end of the line with a soft rubber plug (available at your parts store) to prevent contamination and further loss of fluid. Pull the caliper off the disc **(see illustration)**.

Overhaul

8   Separate the floating frame from the mounting frame **(see illustration)**.
9   Mount the floating frame in a vise so the piston faces down **(see illustration)**.
10   With a soft metal drift, drive the caliper body off the floating frame by striking at both sides of the cylinder **(see illustration)**. Make sure the cylinder comes of squarely. Remove the floating frame from the vise and set it aside.
11   Detach the spring guide from the caliper body **(see illustration)**.
12   Remove the rubber dust cap from the bleeder screw and remove the screw.
13   Reposition the cylinder in the vise with the piston facing up.
14   Pry the dust seal retainer clip from the cylinder **(see illustration)** and remove the dust seal **(see illustration)**.
15   Pull the piston out of the caliper body **(see illustration)**.
16   Inspect the piston and cylinder surfaces. They should be smooth,

4.3a   Disconnect the brake hydraulic line from the chassis by prying it away from the retainer

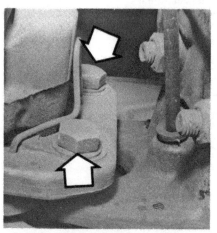

4.3b   Removal of the two caliper retaining bolts will allow the caliper to be removed for servicing or brake disc removal

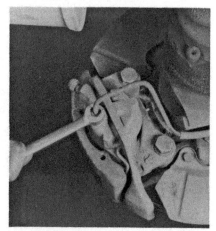

4.4   Use of a flare nut wrench when disconnecting the hydraulic lines will prevent damage to the fittings

4.7   Removing the caliper assembly

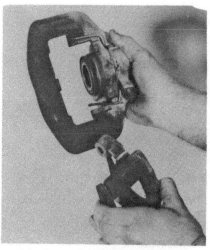

4.8   Separating the mounting frame from the floating frame

4.9   Mount the brake caliper assembly in a vise

with no surface imperfections. *Do not reuse these parts if irregularities are found.*

17 Use a plastic rod, wooden stick or similar tool of soft material to remove the seal from the bore of the caliper (see illustration).

18 With the cylinder repair kit all laid out, lubricate the new piston seal with silicone grease or clean brake fluid. Position the seal in the caliper bore groove, making sure the seal does not twist.

19 Lubricate the piston and caliper bore (see illustrations) with silicone grease or clean brake fluid, set the piston squarely in the bore and push it in until it contacts the seal (see illustration).

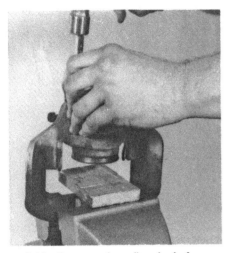

4.10 Separate the caliper body from the frame with a drift and hammer

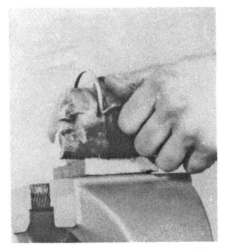

4.11 Remove the caliper spring guide

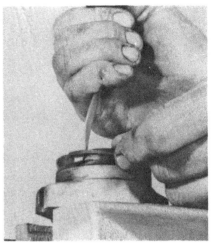

4.14a Pry the retainer clip from the dust seal . . .

4.14b . . . and detach the dust seal

4.15 Removing the piston from the brake cylinder

4.17 Use a wood or plastic tool to remove the caliper seal

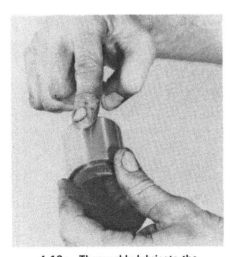

4.19a Thoroughly lubricate the piston . . .

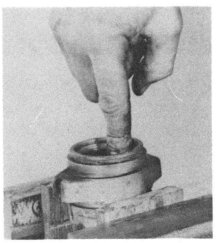

4.19b . . . and the caliper bore

4.19c Push the piston into the cylinder at least as far as the seal, farther if possible

4.20   Installing the dust seal

20 Lubricate and install the dust seal in the groove on the piston (**see illustration**). Locate the cast-in lip of the cylinder and install the clip.
21 Place the cylinder into the vise and push the piston into the cylinder until it is flush with the edge (**see illustration**).
22 Install the bleeder screw and the rubber cap.
23 Install the spring guide.
24 Lightly grease all rubbing surfaces on the caliper body, floating frame and mounting frame. If the rebuild kit provides new plastic slides for the mounting frame, lightly grease and install them (**see illustration**).
25 Slip the guide screw into the groove in the caliper body.
26 Place the mounting frame in the vise and, with a hammer and drift punch, drive the caliper body squarely into the floating frame (**see illustration**). Install the mounting frame on the floating frame.

Installation

27 Installation is the reverse of removal. Install the caliper assembly before installing the brake pads (see Section 3). Be sure to tighten the attaching bolts to the specified torque.
28 Bleed the brake system and clutch (see Section 15).
29 Install the wheel and lower the vehicle. Test brake operation carefully before driving the vehicle in traffic.

4.21   To completely compress the piston, use a vise with soft jaws or small wood blocks as shown

4.24   Replace the plastic slides (if applicable) with the new ones in your rebuild kit

4.26   Be careful when driving the caliper onto the frame. Make sure the caliper bottoms squarely

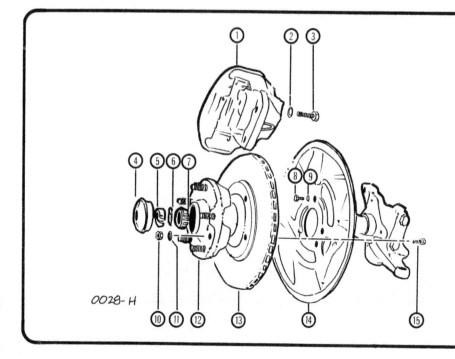

5.1a   Front disc brake assembly — exploded view

1 Caliper assembly
2 Spring washer
3 Caliper retaining bolt
4 Dust cap (late model)
5 Clamping nut with Allen bolt
6 Thrust washer
7 Outer wheel bearing
8 Bolt
9 Washer
10 Hub retaining nut
11 Washer
12 Front wheel hub
13 Brake rotor
14 Guard
15 Rotor retaining bolt

0028-H

5   Brake rotor (disc) — inspection, removal and installation

Refer to illustrations 5.1a, 5.1b, 5.6, 5.7, 5.10a, 5.10b, 5.10c, 5.10d, 5.10e, 5.12, 5.14, 5.16 and 5.18

Inspection

1   Raise the vehicle and support it securely on jackstands. Remove the wheel and hold the rotor in place with two lug nuts **(see illustrations)**. **Note:** *The following information applies to both the front and the rear brakes.*

2   Visually inspect the rotor surface for score marks and damage. Light scratches and shallow grooves are normal after use and are not detrimental to brake operation. Deep score marks, grooves and evidence of overheating (blue areas) require rotor removal and resur-

facing by an automotive machine shop. Be sure to check both sides of the rotor.

3   To check runout, place a dial indicator at a point about 1/2-inch from the outer edge of the rotor. Set the indicator to zero and turn the rotor. The indicator reading should not exceed the specification for installed lateral runout. If it does, the rotor should be resurfaced by an automotive machine shop.

4   Do not use a rotor that has been machined to a thickness less than the minimum. This minimum thickness may be stamped on the rotor itself.

Removal

5   Remove the brake pads (Section 3) and the caliper (Section 4). It will not be necessary to disconnect the hydraulic line unless major work to the caliper is anticipated. Hang the caliper out of the way.

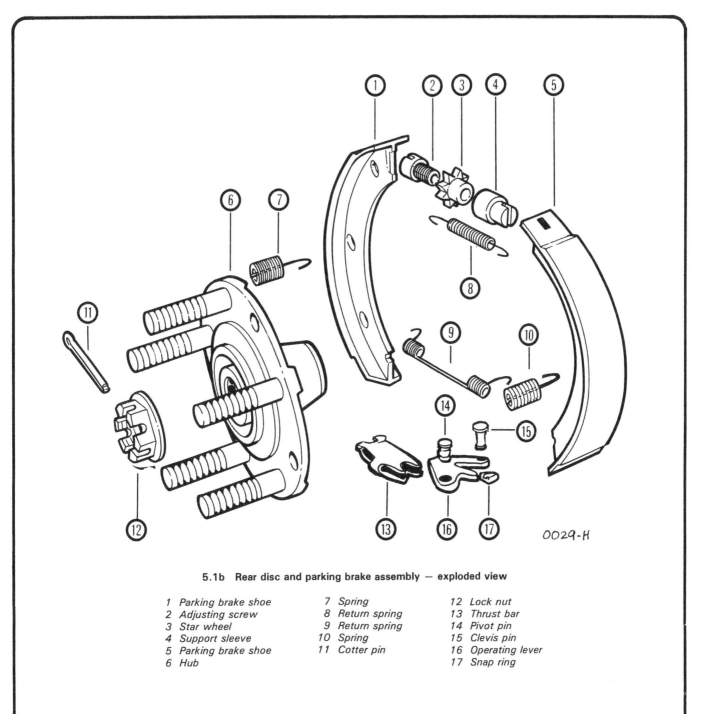

5.1b   Rear disc and parking brake assembly — exploded view

1 Parking brake shoe	7 Spring	12 Lock nut
2 Adjusting screw	8 Return spring	13 Thrust bar
3 Star wheel	9 Return spring	14 Pivot pin
4 Support sleeve	10 Spring	15 Clevis pin
5 Parking brake shoe	11 Cotter pin	16 Operating lever
6 Hub		17 Snap ring

0029-H

5.6   Removal of the front wheel dust cap reveals the clamping nut

5.7   The Allen bolt must be loosened before the clamping nut can be removed

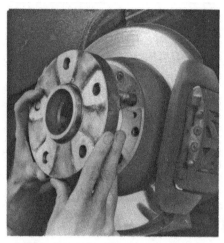

5.10a   Removing the rear wheel spacer

5.10b   Removing the countersunk rotor retaining screw

5.10c   Use a large screwdriver through the adjuster access hole to contract the parking brake shoes so that the rear brake rotor can be removed

5.10d   To free a stuck brake rotor, install two bolts as shown and alternate turns between them . . .

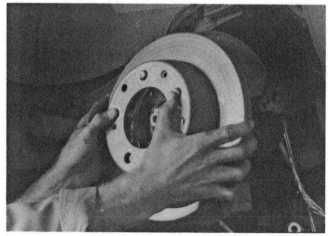

5.10e   . . . until the rotor comes off

Front disc

6   Remove the dust cap by prying it off with a screwdriver **(see illustration)**. The driver's side dust cap on pre-1986 models drives the speedometer cable. This can be easily separated while removing the dust cap.

7   Loosen the threaded clamping nut Allen head bolt **(see illustration)** and remove the nut and thrust washer.

8   Remove the brake rotor by pulling straight out on it. Be careful not to score the spindle or bearing surface.

9   If separating the rotor from the hub is necessary, be sure to mark the rotor and hub so they can be reassembled in the same relative positions. Be sure the rotor-to-hub bolts are correctly tightened during reassembly.

Rear disc

10   The hub on the rear brakes is behind the rotor, so removing it is not necessary. The rear rotors have a thick metal spacer which must be removed before the rotor can come off **(see illustration)**. Once the spacer is removed, remove the countersunk screw from the rotor and pull the rotor from the studs **(see illustration)**. If the rear rotor cannot be removed easily, check to make sure the parking brake is completely released. If the parking brake is released, use a large screwdriver to turn the parking brake star adjuster back until the rotor is free **(see illustration)**. If it is still stuck, thread a couple of bolts into the holes provided and turn each one a little at a time until the rotor comes straight off the studs **(see illustrations)**.

Installation

11   Installation is the reverse of removal. When the operation is complete, lower the vehicle and depress the brake pedal a few times to bring the brake pads into contact with the rotor. Bleeding of the system will not be necessary unless the hose was disconnected from the caliper. Check the operation of the brakes carefully before driving the vehicle. The following points should also be noted.

Front disc

12   Lightly grease the spindle with high-temperature wheel bearing grease before installing the rotor **(see illustration)**. Grease the bearing

5.12 Lubricate the spindle surface with high-temperature wheel bearing grease

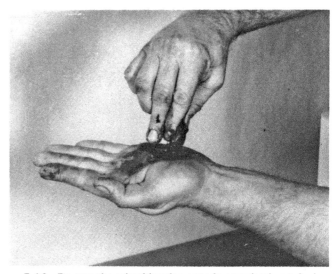

5.14 Be sure the wheel bearing cage is completely packed full of grease before installing the bearing

5.16 Apply lateral force to the thrust washer with a screwdriver to check the clamping nut tightness

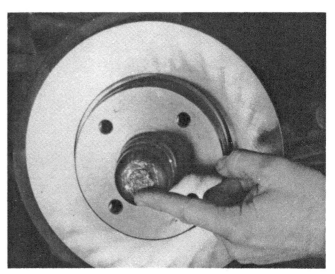

5.18 Hub-driven speedometer cables should be sealed with silicone sealant

surface inside the rotor with the same type of wheel bearing grease.
13 Slide the rotor onto the spindle. A slight resistance should be felt as the seal stretches over the bearing surface.
14 Lightly grease the outer bearing race on the outer wheel bearing with bearing grease. Pack the bearing rollers and cage with bearing grease **(see illustration)**.
15 Install the bearing on the spindle. Slide the thrust washer on behind it, engaging the tang on the washer in the slot in the spindle.
16 Screw on the clamping nut and spin the brake rotor. Continue tightening the nut and spinning the rotor to seat the bearings. When the nut is snug, back it off one-half turn and retighten it until the thrust washer can barely be moved by pressing on it with the blade of a screwdriver **(see illustration)**.
17 Tighten the Allen bolt in the clamping nut to the specified torque.
18 Install the speedometer cable in the dust cap (if applicable) and drive the dust cap into the hub by lightly tapping it with a rubber hammer. Seal the outside of the cable fitting with silicone sealant **(see illustration)**.

Rear disc
19 Place the rotor in position over the threaded studs and install the countersunk screw.
20 Slide the big metal spacer onto the studs and install at least two of the lug nuts to hold the disc/spacer in place while refitting the caliper. Once the caliper is installed (Section 4), remove the lug nuts and install

the wheel.
21 Test the brakes in a safe area before driving the vehicle in traffic.

---

6 Master cylinder — removal, overhaul and installation

Refer to illustrations 6.2a and 6.2b

Note: *Before deciding to overhaul the master cylinder, check on the availability and cost of a new or factory rebuilt unit and also the availability of a rebuild kit.*

Porsche uses master cylinders from two different manufacturers. Be sure you know which one is on your vehicle. If the unit is stamped with a *G*, you've got the Girling version. If the stamp is *Ate*, it's a Teves unit. Another way to identify the make is by the stop screw. The Teves version screw is on the top between the two reservoir inlets. The Girling version is on the side.

Removal

1 Remove the air box to allow clearance for removing the master cylinder. Place rags under the line fittings and prepare caps or plastic bags to cover the ends of the lines once they are disconnected. **Caution:** *Brake fluid will damage paint. Cover all body parts and be careful not to spill fluid during this procedure. If a syringe is available, it may be used to remove the reservoir fluid.*

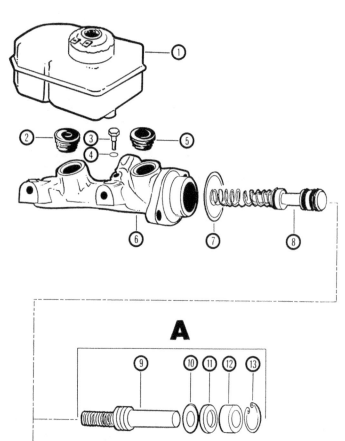

0030-H

6.2a   Brake master cylinder — exploded view (assembly A is Teves version; assembly B is Girling version)

1	Reservoir	11	Secondary cup
2	Grommet	12	Stop washer
3	Stop screw	13	Snap ring
4	Aluminum washer	14	Primary piston
5	Grommet	15	Stop washer
6	Housing	16	Secondary cup
7	Seal	17	Plastic washer
8	Secondary piston	18	Secondary cup
9	Primary piston	19	Stop washer
10	Plastic washer	20	Snap ring

2   Loosen the nuts at the ends of the brake lines where they enter the master cylinder (see illustrations). To prevent rounding off the flats on these nuts, a flare nut wrench, which wraps around the nut, should be used.

3   Pull the brake lines slightly away from the master cylinder and plug the ends to prevent contamination. Detach the electrical connectors from the reservoir cap and separate the reservoir from the master cylinder by gently prying it off with a screwdriver. This will prevent having to disconnect the clutch hydraulic line from the reservoir.

4   Remove the nuts attaching the master cylinder to the power booster. Pull the master cylinder off the studs and out of the engine compartment. Again, be careful not to spill the fluid as this is done.

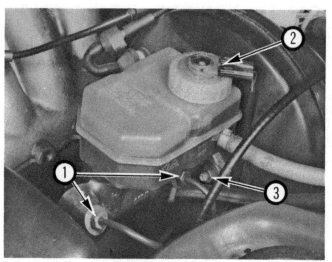

6.2b   Master cylinder

1 Hydraulic line fittings   3 Master cylinder-to-brake
2 Electrical connectors   booster nut (1 of 2)

Overhaul

5   Before attempting the overhaul of the master cylinder, obtain the proper rebuild kit, which will contain the necessary replacement parts and instructions which may be specific to your model.

6   Inspect the reservoir grommets for indications of leakage near the base of the reservoir. If necessary, remove the reservoir cap and diaphragm, inspecting for further damage.

7   Using a wooden dowel or a brass rod, depress the pistons until they bottom out against the other end of the master cylinder. Hold the pistons in this position and remove the stop screw from the top (or side) of the unit.

8   Carefully release the pistons and remove the snap-ring at the end of the master cylinder.

9   The various internal components can now be removed from the cylinder bore. To dislodge the secondary piston assembly you may have to gently bump the open end of the master cylinder against a piece of wood. Make careful note of the installed order of the components so they can be returned to their original locations.

10 Carefully inspect the inside bore of the master cylinder. Any deep scoring or other damage will mean a new master cylinder is required.

11 Carefully note the positions of all components on the pistons, making sketches if necessary. Pay particular attention to the seals and the direction of the seal lips so the new ones can be installed in the same position and direction.

12 Replace all parts included in the rebuild kit, following the instructions in the kit. Clean all re-used parts with clean brake fluid or denatured alcohol. **Warning:** *Do not use gasoline or petroleum-based solvents to clean brake parts.* During assembly, lubricate all parts liberally with clean brake fluid.

13 Use the wood dowel or brass rod to push the assembled components into the bore, bottoming them against the end of the master cylinder, then install the stop screw.

14 Install the new snap-ring, making sure it is seated properly in the groove.

Installation

15 Install the master cylinder over the studs on the power brake booster and tighten the mounting nuts only finger tight at this time.

16 Using your fingers, thread the brake line fittings into the master cylinder. Since the master cylinder is still a bit loose, it can be moved slightly in order for the fittings to thread in easily. Do not strip the threads on the heads as the fittings are tightened (use the flare nut wrench).

17 Tighten the brake fittings and the two mounting nuts.

18 Attach the reservoir securely to the master cylinder with new rubber grommets and attach the electrical connectors to the reservoir cap.

19 Fill the master cylinder reservoir with new, clean brake fluid and bleed the brake system as described in Section 15. Check the brakes carefully before driving the vehicle.

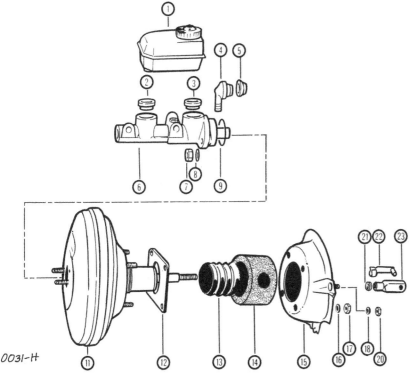

7.1   Brake booster and master cylinder — exploded view

1 Brake fluid reservoir
2 Reservoir plug
3 Reservoir plug
4 Check valve/hose fitting
5 Rubber seal
6 Tandem brake master cylinder
7 Master cylinder mounting nut
8 Washer
9 Seal
11 Brake booster
12 Gasket
13 Dust cover
14 Gasket
15 Adapter
16 Washer
17 Nut
18 Lockwasher
20 Booster adapter mounting nut
21 Pushsrod locknut
22 Lockpin
23 Swivel joint

0031-H

7.6   Removing the brake booster vacuum hose with the check valve installed

7.7   The brake booster pushrod comes through the firewall and attaches to the top of the brake pedal – to disconnect it, loosen the locknut all the way and unscrew the pushrod (it is not necessary to remove the swivel at the pedal)

7.9   Removing the brake booster from the engine compartment

7   Power brake booster — inspection, removal and installation

Refer to illustrations 7.1, 7.6, 7.7 and 7.9

1   The power brake booster unit requires no special maintenance apart from periodic inspection of the vacuum hose and the case **(see illustration)**.
2   Dismantling of the power unit requires special tools and is not ordinarily done by a home mechanic. If a problem develops, a new or factory rebuilt unit should be installed.

Inspection

3   With the engine off, pump the brake pedal several times to dissipate the vacuum in the booster. Depress the pedal and start the engine. If the booster is working properly, there will be a slight drop of the pedal. Medium pressure on the pedal will produce the best results for this test.
4   If the brake booster fails the test, remove it and take it with you when buying the replacement unit. Do not assume that the booster is the same brand as the master cylinder. Porsche uses different com-

binations in the system; you should duplicate the one you have to avoid clearance problems.

Removal

5   Remove the master cylinder (Section 6).
6   Disconnect the vacuum hose where it attaches to the power brake booster **(see illustration)**.
7   From underneath the dashboard in the passenger compartment, disconnect the power brake pushrod from the top of the brake pedal **(see illustration)**.
8   Remove the nuts attaching the booster to the firewall.
9   Carefully lift the booster unit away from the firewall and out of the engine compartment **(see illustration)**.

Installation

10 Place the booster into position and tighten the retaining nuts. Connect the brake pedal.
11 Install the master cylinder and vacuum hose.
12 Carefully test the operation of the brakes before driving the vehicle in traffic.

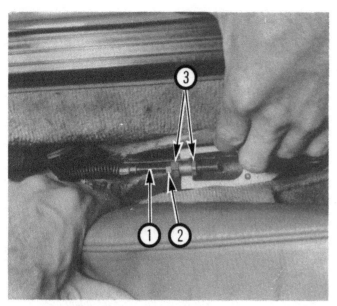

8.3   The threaded end of the parking brake cable (1) goes through the brake handle yoke and is held in place by a locking nut (2) and adjusting nuts (3)

10.5   Before the parking brake shoes can be removed, the adjuster assembly must be pried out

8   Parking brake cable — replacement

Refer to illustration 8.3

1   Raise the vehicle and support it securely on jackstands.
2   Remove the rear wheels and rear brake rotors (Section 5). Disconnect the cable ends from the brake shoe actuating levers, remove the brake cable backing plate and pull the cables through the brake drum backing plate.
3   Remove the driver's seat, pull back the carpeting and release the locking and adjusting nuts from the cable end **(see illustration)**. Remove the cable and unhook the yoke.
4   Crawl under the vehicle and find the place where the cable goes through the body. This should look like a metal plug with the cable housing coming out of the center. Pull sharply to loosen the plug and detach the cable assembly from the body.
5   The two branches of the cable assembly are held in place on the body by two plastic clips. Carefully pry open the clips and remove the remainder of the parking brake cable assembly.
6   Installation of the new cable is the reverse of the removal procedure. Adjust the parking brake as described in Section 9.

9   Parking brake — adjustment

1   Adjusting the parking brake is a two step process.
2   Raise the vehicle and support it securely on jackstands.
3   Pull the parking brake lever two clicks and see if you can turn the rear wheels by hand. You should be able to do this only with a great deal of force. If the brake resistance is incorrect, remove the rear wheels and adjust the parking brake as follows.
4   Release the parking brake and press back on the rear wheel brake pads until the wheels can be turned freely.
5   Insert a screwdriver into the brake adjusting hole in the hub of each rear brake disc and turn the adjusting device until the wheel cannot be turned (see Section 5). Turn the adjuster back just to the point where the wheel can be turned freely.
6   Remove the driver's seat. Pull back the carpeting around the brake lever. You should see a locknut, adjusting nuts and yoke at the end of the cable **(see illustration 8.3)**.
7   Loosen the locknut and adjust the cable tension with the adjusting nuts until the wheels can be just barely turned with the lever on the second click.
8   Tighten the locknut.
9   Replace the carpeting and the seat.
10   Install the wheels and lower the vehicle.

10   Parking brake assembly — check, removal and installation

Refer to illustrations 10.5, 10.7, 10.9, 10.13a, 10.13b, 10.13c and 10.13d
Warning: *Some of the parking brake components produce asbestos dust, which may cause serious harm if inhaled. When servicing these components, do not create dust by grinding or sanding the linings or by using compressed air to blow away the residue. Use a water dampened cloth to wipe away the residue.*

Check

1   With the vehicle parked on a hill, apply the brake, place the transmission in Neutral and see if the parking brake alone will hold the vehicle. This is a simple check that should be performed regularly. However, every 24 months or so (or whenever a fault is suspected), the parking brake assembly itself should be visually inspected.
2   With the vehicle raised and supported on jackstands, remove the rear wheels.
3   Remove the rear rotors as outlined in Section 5. Support the caliper assemblies with a piece of stiff wire and do not disconnect the brake line from the caliper.
4   With the rotor removed, the parking brake components are visible and can be inspected for wear and damage. Most critical are the linings, which can wear over time, especially if the parking brake system has been improperly adjusted. There should be at least 1/8-inch (3 mm) of lining on top of the metal shoes. Also check the springs and adjuster mechanism for damage. Finally, inspect the drum inside of the rotor for deep scratches and other damage.

Removal

5   Spread the shoes apart at the top and remove the adjuster assembly **(see illustration)**.
6   Note how the adjuster return spring is attached to each shoe (it goes on the back of the shoes) and remove the spring with a pair of needle-nose pliers.
7   Using a screwdriver, push down on each hold-down spring and turn it until the hooked end can be removed from the hole in the backing plate **(see illustration)**.
8   Carefully note how the lever and strut assembly (located at the bottom, between the shoes) positions against the forward shoe and the wear shim. Upon installation this assembly must be positioned the same way.
9   With the shoe return spring still attached to the shoes, work the shoes and spring off the backing plate **(see illustration)**.
10   Lift away the shoe assemblies and disengage the shoe return spring

10.7   Press down and turn each hold-down spring to detach it from the brake backing plate

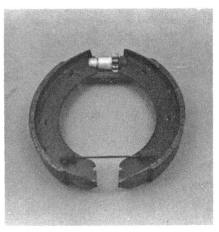

10.9   This is how the parking brake shoes, return springs and adjuster should fit together on the brake backing plate

10.13a   Lubricate all friction bearing surfaces with brake system grease (except the pads), especially the operating lever and strut assembly

10.13b   Install the operating lever . . .

10.13c   . . . the shoes (with bottom return spring attached) . . .

10.13d   . . . and the upper return spring and adjuster assembly

from each shoe. Disassemble the adjuster assembly and inspect for damage. Apply multi-purpose grease to the threads and confirm that the adjuster screw turns easily into the socket.
11   Inspect all parts and replace with new ones if damage is found.

Installation

12   Noting the **Warning** at the beginning of this Section, clean all parts (except the new linings) with denatured alcohol. *Do not use gasoline or petroleum-based solvents to clean brake parts.*
13   Place a thin layer of grease on all areas, except the pads, where friction may occur. Install the return springs in the shoes and work the new shoes into position on the backing plate, making sure the operating lever and strut assembly is positioned correctly **(see illustrations)**.
14   Install the hold-down springs by depressing the springs and inserting the straight portion through the slot in the backing plate. Be sure the hooked end is securely fastened to the backing plate.
15   Spread the shoes and install the adjuster assembly, making sure the star wheel is positioned towards the front.
16   Slide the rotor into position, turning the adjuster as necessary for the rotor to clear the shoes. Install the rotor retaining screw.
17   Install the caliper, referring to Section 4. Install the wheels. Adjust the parking brake as outlined in Section 9.

11   Brake pedal — removal, installation and adjustment

Removal

1   Disconnect the battery ground cable. Place the cable out of the way so that it cannot come into contact with the battery post.
2   Disconnect the brake pedal from the power brake booster pushrod by removing the lockpin. Be careful not to loosen the pushrod locknut.
3   Remove the nut and bolt at the top of the pedal. The brake pedal can now be removed from the bracket.

Installation

4   Install a new bushing if necessary and lubricate it, the pivot bolt and all friction bearing surfaces with heavy duty grease.
5   Place the pedal in position and slide the bolt into place.
6   Tighten the nut and attach the pedal to the booster pushrod.
7   Operate the brake pedal several times to ensure proper operation.

Adjustment

8   Excessive brake pedal travel is an indication of worn brake components (usually disc brake pads) and indicates that a thorough brake system inspection is required.

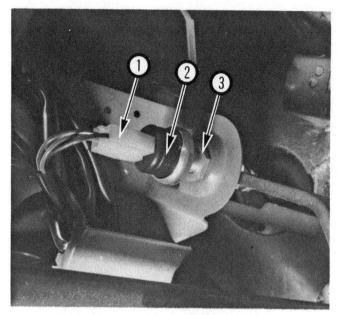

12.1   Brake light switch electrical connector (1),
       switch (2), and locknut (3)

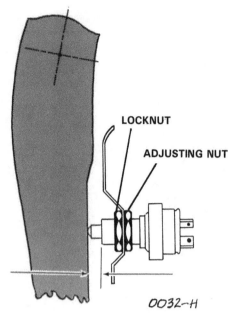

12.6   Measure the distance between the brake light switch
       and the brake pedal

9   The power brake booster can be adjusted at the pushrod, which
attaches to the brake pedal. This adjustment, however, should not be
performed until all remaining brake components have been found to
be in proper working order. Refer to Section 13 for the adjustment
procedure.

12   Brake light switch — removal, installation and adjustment

Refer to illustrations 12.1 and 12.6

Removal

1   The brake light switch is located on a bracket at the top of the
brake pedal **(see illustration)**.
2   Disconnect the negative battery cable and secure it out of the way
so that it cannot come into contact with the battery post.
3   Disconnect the wire harness at the brake light switch.
4   Remove the nut on the pedal side of the bracket. Depress the brake
pedal and remove the switch.

Installation and adjustment

5   With the brake pedal depressed, install the new switch and the
locknut.
6   To adjust the brake light switch, use the adjusting nut and locknut
to control the distance between the brake pedal and the switch **(see
illustration)**. The correct distance is 1/4-in (6 mm). When this distance
is correct, tighten the locknut.
7   Connect the wiring at the switch and the battery. Make sure that
the rear brake lights are functioning properly.

13   Brake pushrod — adjustment

1   The brake pushrod does not require adjustment unless the brake
booster is replaced, the pivot end of the brake pushrod has been removed, or the pushrod or pivot end has been turned.
2   The brake pushrod is adjusted by turning the pivot end, located
under the dash and connected to the top of the brake pedal assembly.
Use a pair of pliers to remove the lockpin.
3   After removing the pin, the pushrod can be detached from the pedal

assembly. Loosen the locknut and turn the pivot end in or out until
the correct length is attained.
4   When the pushrod is properly adjusted, the distance from the surface of the brake booster mount to the center of the lockpin on the
pivot end should be 7.25-inches (186 mm).
5   When the distance is correct, tighten the locknut and install the
lockpin.
6   Test the brakes in a safe area before driving the vehicle in traffic.

14   Brake hoses and lines — inspection and replacement

Refer to illustration 14.3

1   About every six months, with the vehicle raised and supported
securely on jackstands, the flexible hoses which connect the steel brake
lines with the front and rear brake calipers should be inspected for
cracks, chafing of the outer cover, leaks, blisters and other damage.
These are important and vulnerable parts of the brake system and
inspection should be complete. A light and mirror will prove helpful
for a thorough check. If a hose exhibits any of the above conditions,
replace it with a new one.

Flexible hose replacement

2   Clean all dirt away from the ends of the hose.
3   Disconnect the brake line from the hose fittings using a back-up
wrench on the fitting **(see illustration)**. Be careful not to bend the frame
bracket or line. If necessary, soak the connections with penetrating oil.
4   Remove the U-clip from the female fitting at the bracket and remove
the hose from the bracket.
5   Disconnect the hose from the caliper (front wheels).
6   Attach the new brake hose to the caliper.
7   Pass the hose through the frame bracket (front wheels). With the
least amount of twist in the hose, install the fitting.
8   Install the U-clip in the female fitting at the frame bracket.
9   Attach the brake line to the hose fittings using a back-up wrench
on the fitting.
10   Carefully check to make sure the suspension or steering components do not make contact with the hoses. Have an assistant push
on the vehicle and also turn the steering wheel from lock-to-lock during
the inspection.
11   Bleed the brake system as described in Section 15.

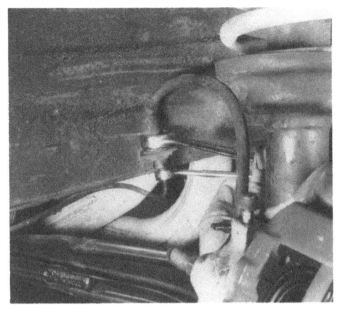

14.3   Use a backup wrench on the hydraulic line fittings (front caliper shown)

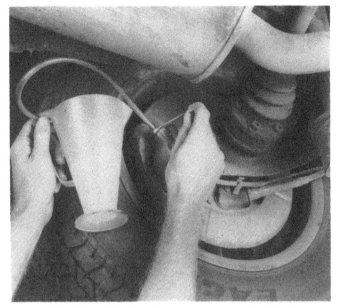

15.9   When bleeding the rear brakes, get the bleeder screw as close as possible to the 11 o'clock position to keep air from entering the system

Rigid (metal) brake line replacement

12 When replacing brake lines do not use copper tubing for any brake system connections. Purchase steel brake line from a dealer or auto parts store.
13 Prefabricated brake lines, with the ends already flared and fittings installed, are available. These lines are also bent to the proper shapes.
14 If prefabricated lines are not available, obtain the recommended steel tubing and fittings to match the line to be replaced. Determine the correct length by measuring the old brake line (a piece of string can usually be used for this) and cut the new tubing to length, allowing about 1/2-inch extra for flaring the ends.
15 Install the fitting over the cut tubing and flare the ends of the line.
16 If necessary, carefully bend the line to the proper shape. A tube bender is recommended for this. **Warning:** *Do not crimp or damage the line.*
17 When installing the new line make sure it is well supported in the brackets and has plenty of clearance between moving or hot components.
18 After installation, check the master cylinder fluid level and add fluid as necessary. Bleed the brake system as outlined in the next Section and test the brakes carefully before driving the vehicle in traffic.

---

15   Brake system and clutch — bleeding

Refer to illustration 15.9

Note: *Bleeding the hydraulic system is necessary to remove any air that manages to find its way into the system when it has been opened during removal and installation of a hose, line, caliper, clutch slave cylinder or master cylinder.*

1 It will probably be necessary to bleed the system at all four brakes and the clutch if air has entered the system due to low fluid level, or if the brake lines have been disconnected at the master cylinder.
2 If a brake line was disconnected at only one wheel, then only that caliper must be bled.
3 If a brake line is disconnected at a fitting located between the master cylinder and any of the brakes, that part of the system served by the disconnected line must be bled.
4 If the rear brakes are being bled, raise the front of the vehicle, which will position the bleeder valve at the 11 o'clock position and prevent air from being trapped in the caliper.
5 Remove any residual vacuum from the brake power booster by

applying the brake several times with the engine off.
6 Remove the master cylinder reservoir cover and fill the reservoir with brake fluid. Reinstall the cover. **Note:** *Check the fluid level often during the bleeding operation and add fluid as necessary to prevent the fluid level from falling low enough to allow air bubbles into the master cylinder.*
7 Have an assistant on hand, as well as a supply of new brake fluid, an empty clear plastic container, a length of 3/16-inch plastic, rubber or vinyl tubing to fit over the bleeder screw and a wrench to open and close the bleeder screw.
8 Beginning at the left rear wheel, loosen the bleeder screw slightly, then tighten it to a point where it is snug but can still be loosened quickly and easily.
9 Place one end of the tubing over the bleeder screw fitting and submerge the other end in brake fluid in the container **(see illustration)**.
10 Have the assistant pump the brakes a few times to get pressure in the system, then hold the pedal firmly depressed.
11 While the pedal is held depressed, open the bleeder screw just enough to allow a flow of fluid to leave the valve. Watch for air bubbles to exit the submerged end of the tube. When the fluid flow slows after a couple of seconds, tighten the screw and have the assistant release the pedal.
12 Repeat Steps 10 and 11 until no more air is seen leaving the tube, then tighten the bleeder screw and proceed to the right rear wheel, the left front wheel and the right front wheel, in that order, and perform the same procedure. Be sure to check the fluid in the master cylinder reservoir frequently.
13 To bleed the clutch, follow the same procedure with the bleeder screw on the back of the clutch slave cylinder, having your assistant pump the clutch pedal instead of the brake pedal.
14 Never re-use old brake fluid. It contains moisture which will deteriorate the brake system components.
15 Refill the master cylinder with fluid at the end of the operation.
16 Check the operation of the brakes. The pedal should feel solid when depressed, with no sponginess. If necessary, repeat the entire process. **Warning:** *Do not operate the vehicle if you are in doubt about the effectiveness of the brake system.*
17 If any difficulty is experienced in bleeding the hydraulic system, or if an assistant is not available, a pressure bleeding kit is a worthwhile investment. If connected in accordance with the instructions, each bleeder screw can be opened in turn to allow the fluid to be pressure ejected until it is clear of air bubbles without the need to replenish the master cylinder reservoir during the process.

Chapter 10 Steering and suspension systems

Contents

Specifications

Front suspension

Torque specifications	Ft-lbs	Nm
Control arm-to-crossmember bolt/nut	48	65
Control arm-to-body bolts	34	46
Aluminum control arm-to-control arm mount	63	85
Control arm-to-steering knuckle clamp bolt/nut	37	50
Crossmember-to-body	63	85
Stabilizer mount-to-body	17	23
Stabilizer clamp-to-mount	17	23
Stabilizer-to-steel control arm	17	23
Stabilizer mount-to-aluminum control arm	18	25
MacPherson strut-to-upper mount nuts	25	34
MacPherson strut-to-steering knuckle bolts/nuts	74	100
Balljoint mounting nut	10 to 12	13 to 16
Splash guard-to-steering knuckle	7.5	10
Caliper-to-steering knuckle	63	85
Brake disc-to-wheel hub bolts	17	23
Guide joint-to-steel control arm	19	25

Rear suspension

Torque specifications	Ft-lbs	Nm
Bearing flange-to-cross tube	34	46
Bearing flange-to-body	52	70
Thrust bearing-to-bearing flange	34	46
Thrust bearing-to-body	34	46
Mount-to-body	34	46
Mount-to-strut	17	23
Trailing arm-to-rear axle strut	66	90
Trailing arm-to-cross tube	45	61
Shock absorber-to-body bolt/nut	45	61
Shock absorber-to-steel trailing arm bolt/nut	45	61
Shock absorber-to-aluminum trailing arm bolt/nut	90	123
Stabilizer suspension-to-rear axle strut and stabilizer	34	46
Stabilizer clamp-to-rear axle cross tube	17	23
Brake backing plate-to-steel trailing arm	43	52
Wheel hub-to-stub axle (steel control arm)	280 to 332	338 to 450
Wheel hub-to-stub axle (aluminum control arm	340	460

Steering system

Torque specifications	Ft-lbs	Nm
Steering intermediate shaft bolt	22 to 26	30 to 35
Steering gear-to-crossmember	17	23
Tie-rod-to-steering knuckle nut	22 to 36	30 to 50
Tie-rod-to-rack (non-power steering only)	36	50
Tie-rod joint-to-tie rod (non-power steering only)	22	30
Steering wheel-to-steering column nut	33	45
Power steering tie-rod-to-rack	52	70
Power steering tie-rod joint-to-tie-rod	52	70
Steering gear feed and return lines	15	20
Power steering pump feed line	22	30
Power steering pump intake hose (hollow bolt)	33	45

1 General information

This Chapter is composed of three sections: front suspension, rear suspension and steering system.

The front suspension is a MacPherson strut type. The lower end of the strut (shock absorber) attaches to a control arm. Post 1985-1/2 models have cast aluminum alloy control arms. There are few rebuildable parts on the front suspension; most of the work in this area will be replacement tasks.

The rear suspension consists of a control arm (also known as a trailing arm) which is connected to the wheel hub at one end and to the torsion bar at the other. Torsion bar response is damped by a shock absorber. Power is delivered to the rear wheels through axles with constant velocity (CV) joints on either end.

The rear suspension has specific settings for caster and camber. Whenever you adjust the ride height or remove a major component from the rear suspension (except the shock absorbers), have the alignment checked by a professional.

Rack and pinon steering is standard. Power steering was offered as an option.

2 Front hub — removal and installation

Refer to illustration 2.4

1   Raise the front of the vehicle, support it securely on jackstands and remove the front wheel.

2   Remove the brake caliper assembly (Chapter 9) and hang it out of the way on a piece of wire so there is no strain on the brake hose.

3   Remove the brake rotor (Chapter 9).

4   Remove the bolts and nuts and detach the hub and bearing assembly from the disc (see illustration).

5   Installation of the hub to the disc is the reverse of removal. Be sure to tighten the bolts to the specified torque.

6   Install the brake rotor and adjust the bearing play as outlined in Chapter 1.

7   Install the caliper and wheel and lower the vehicle. Road test the vehicle before returning it to normal operation.

2.4 Separating the hub from the disc must be done on a bench (it is crucial that the hub retaining bolts be properly tightened during reassembly)

3   Front stabilizer bar — removal and installation

Refer to illustrations 3.3a and 3.3b

1   Raise the front of the vehicle, support it securely on jackstands and remove the front wheels. Apply the parking brake.
2   Remove the seven bolts and detach the undercover. The stabilizer bar is held in place by two clamps which hold it to the body and two end pieces which attach it to the control arms.
3   Remove the two stabilizer bar clamp bolts from the body **(see illustration)** and the two bushing retainers at the control arms **(see illustration).**
4   Remove the stabilizer bar from the vehicle.
5   To install, place the stabilizer bar in position on the frame rails and install the clamps and bolts.
6   Attach the control arm ends to the control arms and tighten the bolts to the specified torque.
7   Install the wheels, lower the vehicle weight onto the suspension and tighten the stabilizer clamp bolts to the specified torque.

4   MacPherson strut — removal and installation

Refer to illustrations 4.2, 4.5 and 4.6

1   Disconnect the negative cable from the battery.
2   Loosen each of the strut mounting nuts in the engine compartment about two turns **(see illustration)**. Do not loosen them more than two turns and do not remove them.
3   Raise the vehicle and place it securely on jackstands.
4   Remove the front wheels.
5   Pull the rubber brake hose from the clip on the strut **(see illustration).**

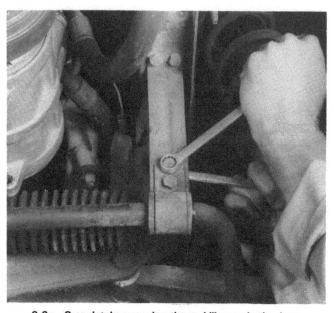

3.3a   Completely removing the stabilizer-to-body clamp bolts is only necessary when removing the stabilizer bar (they should only be loosened when servicing the control arms)

3.3b   Removing the stabilizer-to-control arm bushing retainer mounting bolts is necessary both for control arm and stabilizer bar removal

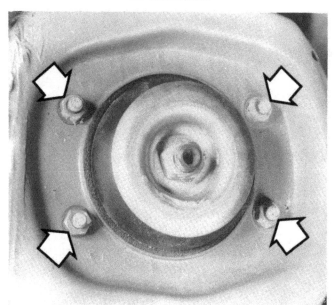

4.2   The MacPherson strut upper mounting nuts (arrows) are accessible from under the hood — do not, under any circumstances, loosen or remove the large center nut

4.5   Release the rubber brake hose from the clip before removing the strut

6   The upper bolt on the strut mounting plate is an eccentric cam. Scribe or paint a thin line on the cam and mounting plate for reassembly purposes **(see illustration)**. This will give you a general alignment setting.

7   Remove the self-locking nuts and bolts from the strut mounting plate. Discard the self-locking nuts and replace them with new ones when installing the strut. Pull the strut free from the steering knuckle.

8   Remove the upper mounting nuts and lower the strut out of the body.

9   Installation is the reverse of removal. Locate the four mounting studs in the holes in the wheel well and install the nuts. Tighten the nuts about two turns each in a criss-cross pattern until the nuts are snug and the upper mount is firmly in place. At this time, do not tighten the nuts to the specified torque — leave them loose enough to allow for some give in the plate.

10   Slide the mounting plate onto the steering knuckle and install the new self-locking nuts finger tight. Align the mark you made on the eccentric cam with the mark on the plate and tighten the nut to the specified torque.

11   Tighten the upper mount nuts to the specified torque.

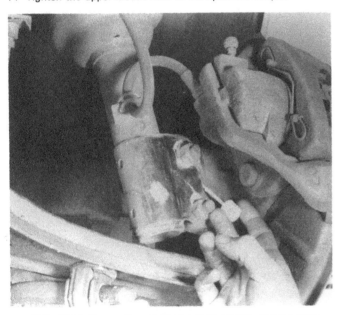

4.6   **Mark the position of the mounting bolts and the strut mounting plate before removing the strut**

12   Replace the wheel and lower the vehicle.

13   Take the vehicle as soon as possible to an alignment shop for final adjustment.

5 MacPherson strut — overhaul

Disassembly and overhaul of MacPherson struts require specialized equipment and training. *Do not attempt to service the strut yourself.*

6 Shock absorbers — inspection, removal and installation

Refer to illustrations 6.6 and 6.7

Inspection

1   The simplest test of the shock absorber's condition is to depress the rear corners of the vehicle several times and note whether the vehicle stops bouncing when you stop moving it. If the vehicle rebounds more than twice, the shocks should be replaced.

2   If the shocks pass the bounce test, crawl under the vehicle and inspect them for fluid leakage, punctures and deep dents. Also inspect the shock body for signs of friction due to the unit being bent. Oil on the outside of the shock absorber is an indication that the seals have started to leak and the units must be replaced. If the shock absorber has failed internally, the vehicle ride will be affected, particularly over uneven surfaces. If a shock absorber is suspected to have failed, remove it from the vehicle and, holding it in a vertical position, extend and collapse it through the full length of travel several times. Any lack of resistance in either direction will indicate the need for replacement. Always replace shock absorbers in pairs.

Removal and installation

3   Procedures for removing the MacPherson Strut (front shock absorber) can be found in Section 4 of this Chapter.

4   Raise the vehicle, support it securely on jackstands and remove the rear wheels.

5   Place a jack beneath the control arm directly under the shock absorber. Do not raise the jack yet.

6   Remove the nut from the lower mounting bolt of the shock absorber **(see illustration)**. Raise the control arm until the lower mounting bolt can be removed from the hole.

7   Compress the shock absorber and remove the nut and bolt from the upper shock absorber mount **(see illustration)**. Remove the shock absorber by lifting it out from between the mounts.

6.6   **Remove the nut from the rear shock absorber lower mounting bolt**

6.7   **The nut on the rear shock absorber upper mounting bolt is located inside the wheel well**

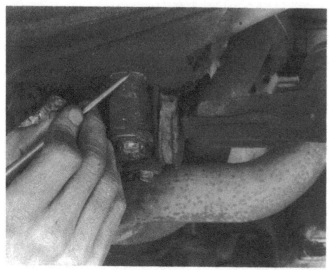

7.3a   Scribe a line around the perimeter of the control arm inner rear mount

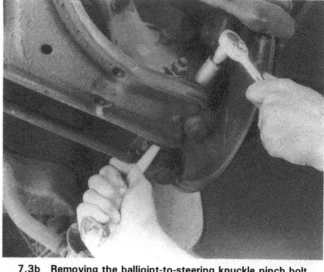

7.3b   Removing the balljoint-to-steering knuckle pinch bolt will allow separation of the two components for servicing of the control arm, balljoint or steering knuckle

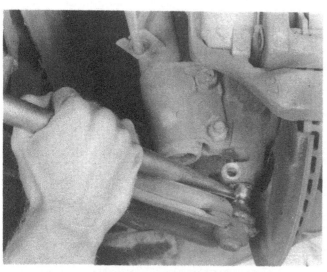

7.3c   Separate the the control arm from the steering knuckle with a pry bar

7.4a   Remove the front control arm forward pivot bolt

8   Installation is the reverse of removal. Tighten the mounting nuts to the specified torque.

7   Front control arm — removal and installation

Refer to illustrations 7.3a, 7.3b, 7.3c, 7.4a, 7.4b and 7.5

1   Raise the vehicle, support it securely on jackstands and remove the front wheel.
2   Disconnect the stabilizer bar from the steering knuckle and let it hang from the body.
3   Scribe a line around the control arm rear inner mount bracket to aid in alignment upon reassembly **(see illustration)**. Disconnect the control arm from the steering knuckle by removing the pinch bolt and separating the two pieces with a pry bar or large screwdriver **(see illustrations)**.
4   Remove the bolt and self-locking nut from the forward, inner mount **(see illustration)**. On some models it may be necessary to disconnect the steering rack clamps to provide clearance to remove the front pivot bolt **(see illustration)**. Discard the self-locking nut and replace it with a new one upon reassembly.

7.4b   If extra clearance is necessary to remove the forward control arm pivot bolt, loosen the steering rack mounting clamps

7.5 Removing the control arm rear inner mount bracket bolts will allow the control arm to be removed from the vehicle (be sure the exact position of the mount has been scribed on the frame)

8.6 Drill diameter should be slightly smaller than the diameter of the bolts included with the new balljoint — do not allow the drill bit to contact the control arm

5   Remove the rear inner mount bracket bolts **(see illustration)** and detach the control arm from the vehicle. If the nut is a self-locking type, replace it with a new one. Some versions may use a nut with a lock washer instead.
6   Replacing the control arm bushings in steel control arms requires a hydraulic press. We recommend that the unit be taken to a repair shop.
7   If your vehicle has cast aluminum alloy control arms, neither bushing, balljoint nor rubber boot replacement is possible. The control arm must be replaced as a unit.
8   Installation is the reverse of removal. Be sure to use new self-locking nuts. Tighten all fasteners to the specified torque.
9   Have the front end alignment checked by a dealer service department or an alignment shop.

8 Balljoints — check and replacement

Refer to illustration 8.6
Check
1   Unlock the steering lock, raise the front of the vehicle and support it securely under the control arms on jackstands placed as close as possible to the balljoints.
2   Check the protective rubber caps by pulling them back with your fingers. Look for cracks and hardening of the rubber.
3   To check balljoint play, follow the procedure in Chapter 1.

Replacement
4   Raise the vehicle, support it securely on jackstands and remove the front wheels.
5   Remove the control arm (Section 7) and hold it securely in a soft jawed vise with the pivot end of the balljoint facing out.
6   Drill through the rivets with a drill bit slightly smaller in diameter than the bolts that come with the replacement balljoint **(see illustration)**. Drill only as far as necessary to allow the rivet to be knocked out with a punch and hammer. Don't drill into the control arm. Inspect the holes in the steering knuckle and control arm, removing any accumulated dirt. If out-of-roundness, deformation or other damage is noted, the knuckle (Section 15) or control arm must be replaced.
7   Drive out the remaining rivet pieces with a soft metal drift and remove the old balljoint.

9.2a Discard the cotter pin after removing it from the hub nut

8   Making sure all components are completely clean, install the new balljoint. Install the mounting bolts for the new balljoint from the top and install the nuts on the bottom. Tighten the nuts to the specified torque.
9   Install the control arm and lubricate the new balljoints.
10 Install the wheels and lower the vehicle.
11 The front end alignment should be checked by a dealer service department or alignment shop.

9 Rear hub — removal and installation

Refer to illustrations 9.2a, 9.2b and 9.3
1   Raise the rear end, secure the vehicle on jackstands and remove the wheel.
2   Remove the spacer (if applicable) and brake disc (Chapter 9). The cotter pin comes out before the nut comes off **(see illustrations)**. Models with aluminum trailing arms don't have a cotter pin.

9.2b   Removing the hub nut may require the use of a pry bar as shown to prevent the hub assembly from turning

9.3   Removing the rear hub is usually possible without the use of a puller (lightly coat the axle splines with moly-base grease during reassembly)

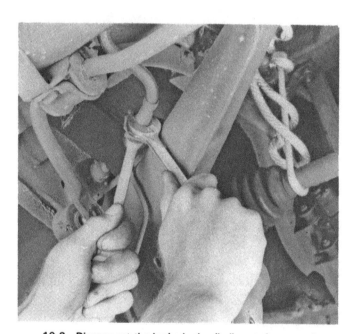

10.3   Disconnect the brake hydraulic line at the control arm mount before removing the control arm

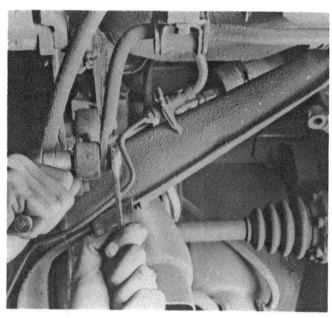

10.5   Disconnect the rear stabilizer bar from the control arm link

3   It is unlikely that removing the hub will require considerable force. Use of a gear puller is recommended if it is stuck **(see illustration)**.

4   Check that the splines in the hub and on the axle are in good condition. Lubricate the splines with a light coat of moly-base grease before installing the hub.

5   When installing the hub, use a big socket to drive the hub as far onto the stub axle as it will go, then use a heavy-duty torque wrench that will measure up to 340 ft-lbs (460 Nm) to tighten the hub nut. Align the cotter pin hole with the hole in the axle by tightening the nut, never by backing it off.

6   Install a new cotter pin.

7   Install the brake disc and spacer.

8   Install the wheel and lower the vehicle.

10   Rear control arm — removal, overhaul and installation

Refer to illustrations 10.3, 10.5, 10.6, 10.7, 10.10, 10.12, 10.14, 10.15, 10.16, 10.17 and 10.26

Removal

1   Disconnect the negative cable from the battery.

2   Raise the rear of the vehicle and remove the wheels.

3   After removing the brake rotor and brake backing plate (Chapter 9), disconnect the brake hydraulic line from the tab on the control arm **(see illustration)**.

4   Remove the hub (Section 9) and disconnect the parking brake cable by prying off the small clip at the base of the parking brake assembly. This will allow the parking brake cable housing to be disconnected from

10.6 Never allow the axle/CV joint assembly to hang freely from the transmission, as it can damage the joint or cause it to come apart (keep it as horizontal as possible)

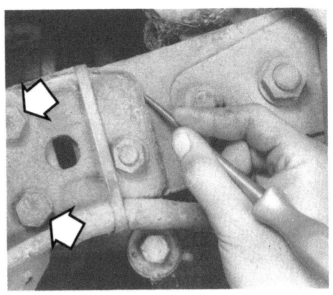

10.7 Scribe a line around the control arm mounting plate as shown (when disconnecting the control arm from the plate, first remove the two vertically aligned bolts [arrows], then the remaining one)

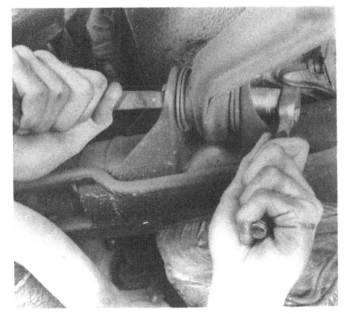

10.10 Remove the rear control arm inner pivot bolt/nut

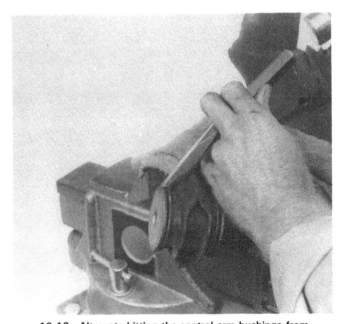

10.12 Alternate hitting the control arm bushings from side-to-side so they come squarely out of the bushing cavity

behind the parking brake shoe assembly.

5 Disconnect the stabilizer bar (see illustration).

6 Detach the driveaxle from the stub axle (Chapter 8), suspend the axle with rope or wire (see illustration) to prevent damage to the in-board CV joint and remove the stub axle (Section 11).

7 Scribe a line around the edge of the control arm mounting plate to aid in accurate reassembly. Of the three bolts holding the control arm to the spring plate, remove the two which align vertically. Loosen, but do not remove, the third one (see illustration).

8 Remove the lower shock absorber mounting bolt and compress the shock absorber.

9 Aluminum control arms come equipped with a vibration damper. Disconnect the vibration damper from the control arm.

10 Remove the self-locking nut from the inner control arm pivot bolt (see illustration) and replace it with a new one upon reassembly.

11 Support the control arm with one hand and remove the bolts from the spring plate and the inner pivot with the other. Remove the control arm from the vehicle and take it to your work area for inspection and servicing.

Overhaul

Steel control arms

12 Clamp the control arm in a vise and remove the two rubber bushings from the forward end of the control arm by driving them out with a cold chisel and hammer (see illustration). Apply leverage on alternate sides of the bushings so they come out squarely.

13 Remove the inner spacer.

14 Pry the inner seal out of the control arm with a large screwdriver **(see illustration)**.
15 Remove the snap-ring. The bearings may be driven out from the inboard side with a soft drift and hammer. Strike the bearing races on opposite sides to ''walk'' the races out **(see illustration)**.
16 Remove the rest of the components from the bearing cavity and clean them and the cavity thoroughly. If using water based degreasing solvents, be sure to let the control arm dry completely before reassembling it **(see illustration)**.
17 Assembly is the reverse of the above procedure. Using a bearing driver, appropriately sized socket or soft-face hammer, drive the inboard bearing all the way to its stop inside the control arm **(see illustration)**. The chamfered edge of the bearing outer race should face out.
18 Install the snap-ring. Replace it with a new one if it isn't perfectly flat.
19 Turn the control arm over and pack the bearing cavity with approximately 80 grams (3 ounces) of multi-purpose grease.
20 Install the spacer in the cavity and locate it in about the center of the inner bearing.
21 Install the outboard bearing with the chamfered edge facing out.
22 In the absence of a hydraulic bearing press, use a vise and a pair of wood blocks to press the inboard and outboard bearings together until they are flush with the surface of the control arm.
23 Install the inboard seal with a seal driver or large socket.
24 Install the inner spacer and the outboard bearing inner race.
25 Install the stub axle either with a hydraulic press or by driving it squarely into the control arm with a hammer and wood block.
26 Replace the rubber bushings in the front of the control arm by driving them in with a hammer and then squeezing them with a vise until the bushing flanges are tight against the control arm **(see illustration)**.

Aluminum control arms
27 To replace only the rubber bushings on the leading end of the control arm, follow the above procedure.
28 Replacement of the bearings in the trailing end of the aluminum control arm requires special equipment and should be done by a dealer service department.

Installation

29 Installation is the reverse of removal.
30 Align the control arm mounting plate with the line on the spring arm. Do this after you have installed the inner pivot bolt, the lower shock absorber mounting bolt, and pressed the three spring plate bolts into their holes. Do not install any nuts on the bolts.

31 When the single forward bolt on the plate is installed and the mounting plate aligns with the scribed mark, thread a nut onto the single bolt and tighten it enough to hold the pieces in their proper relationship to each other.
32 Install the nuts on the other two mounting plate bolts and the nut on the lower shock absorber mounting bolt.
33 Install the new locknut on the inner pivot bolt and tighten it until it is snug.
34 Tighten the nuts on the three mounting plate bolts to the specified torque, then tighten the lower shock absorber mounting bolt.
35 Install the driveaxle bolts (Chapter 8).
36 Install the brake rotor and the wheel and lower the vehicle so its entire weight is on the wheels.
37 Tighten the pivot bolt locknut to the specified torque.
38 Take the vehicle to a dealer or alignment shop to have the rear suspension aligned.

---

11   Rear wheel stub axle — removal and installation

Refer to illustration 11.3

1 Remove the rear hub (Section 9).
2 Remove the driveaxle and CV joint assembly (Chapter 8).
3 The axle can now be removed through the backside of the control arm. Push it out with a gear puller **(see illustration)** or a hammer and wood block.
4 Installation is the reverse of removal. Be sure to carefully follow the procedures in this Chapter when installing the rear hub.
5 See Section 10 for procedures on servicing the bearings in the control arm.

---

12   Rear stabilizer bar — removal and installation

1 Raise the vehicle and support it securely on jackstands.
2 Disconnect the stabilizer bar at the end links **(see illustration 10.5)**, remove the clamp nuts from the body brackets and detach the bar from the vehicle.
3 To install, place the bar in position and install the brackets and nuts. Tighten the nuts to the specified torque.

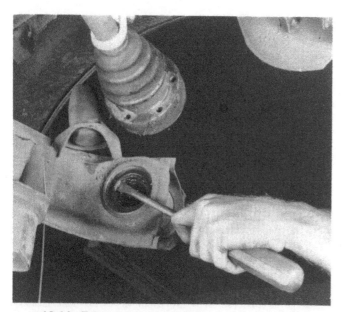

10.14   Take care not to score the grease seal mating surface when prying the seal out of the bearing cavity

10.15   Drive the inboard bearing out as shown, tapping around the edge of the bearing race to bring it squarely out of the cavity (drive the outboard bearing out through the other side using the same method)

4   Connect the bar ends to the links and attach them to the spring plate. Raise each control arm to normal ride height with a jack before tightening the link bolts and nuts to the specified torque.
5   Install the wheels and lower the vehicle.

13 Steering system — general information

All models are equipped with rack-and-pinion steering. Optional power assisted steering was available. The steering gear operates the steering arms via tie-rods, the inner ends of which are protected by rubber boots. Inspect the boots periodically for secure attachment, tears and leaking lubricant.

The power assist system consists of a belt driven pump and asso-

ciated lines and hoses. The power steering pump reservoir fluid level should be checked periodically (Chapter 1).

The steering wheel operates the steering shaft, which actuates the steering gear through universal joints on the intermediate shaft. Looseness in the steering can be caused by wear in the steering shaft universal joint, the steering gear, the tie-rod ends and loose mounting bolts.

14 Steering tie-rod ends — removal and installation

Refer to illustrations 14.2, 14.3 and 14.4

1   Raise the front of the vehicle, support it securely, block the rear wheels and set the parking brake. Remove the front wheels.

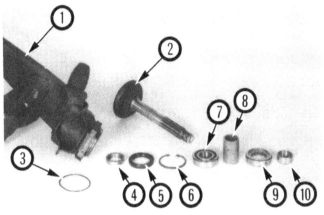

10.16   Components of the steel rear control arm

1 Control arm	6 Snap-ring
2 Stub axle	7 Inboard ball bearing
3 O-ring	8 Spacer
4 Inboard bearing inner race	9 Outboard bearing
	10 Outboard bearing inner race
5 Seal	

10.17   Bearings on both sides may be driven in with a soft-face hammer if a bearing driver is not available

10.26   The control arm inner pivot bushings should be pressed into the control arm with a vise (a light coat of detergent should be used, if necessary, to lubricate the bushings for installation)

11.3   The best method for removing the stub axle is to use a gear puller as shown — a less desirable option is to knock the stub axle out with a hammer using a wood block as a cushion

14.2   The tie-rod end jam nut should be loosened enough to paint an alignment mark on the threads

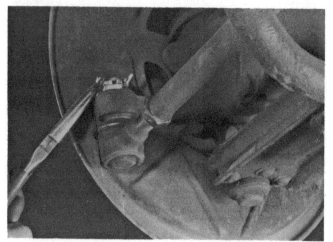

14.3   Discard the cotter pin and replace it with a new one — when aligning the slot in the nut with the cotter pin hole, always tighten the nut rather than loosening it

2   Loosen the jam nut on the end of the tie-rod sufficiently to mark the position of the tie-rod end on the tie-rod threads **(see illustration)**.
3   Remove the cotter pin and loosen the castellated nut on the ball-joint stud **(see illustration)**.
4   Disconnect the tie-rod from the steering knuckle arm with a puller or "pickle fork" **(see illustration)**.
5   Installation is the reverse of removal. Thread the tie-rod end onto the rod to the marked position and tighten the jam nut securely.
6   Connect the tie-rod end to the steering knuckle arm, install the nut and tighten it to the specified torque. Install a new cotter pin.
7   Have the front end steering alignment checked by a dealer service department or an alignment shop.

15   Steering knuckle — removal, inspection and installation

1   Raise the vehicle, support it securely on jackstands and remove the front wheels.
2   Remove the brake caliper mounting bolts and hang the caliper out of the way on a piece of wire. Do not allow the caliper to hang by the brake hose.
3   Remove the hub and disc assembly (Section 2).
4   Unbolt and remove the splash shield.
5   Disconnect the tie-rod end (Section 14)
6   The upper mounting bolt on the MacPherson strut is an eccentric cam. Paint or scribe a line across the cam and the strut mounting plate on both the bolt head and nut sides of the plate **(see illustration 4.6)**. Remove the nuts from the strut mounting bolts, but do not remove the bolts. Discard the nuts and replace both of them with new ones upon installation.
7   Remove the clamping bolt and nut from the control arm (balljoint) mount. Discard of the self-locking nut and replace it with a new one upon reassembly.
8   Remove the bolts form the strut mounting plate and remove the steering knuckle from the vehicle.
9   Inspect the steering knuckle for damage and signs of stress or wear on the wheel spindle.
10   Installation is the reverse of removal. Tighten all fasteners to the specified torque.
11   Remember to use new self-locking nuts and to set the eccentric cam back in the marked locations. Aligning the eccentric cam adjusters with their previous marks will give you basic camber and toe settings sufficient to get you to an alignment shop for final adjustment. Do not put the vehicle into normal operation until the front wheels have been correctly aligned.

16   Steering gear — removal and installation

Refer to illustrations 16.7 and 16.8
1   Disconnect the negative cable from the battery, raise the vehicle

and secure it on jackstands.
2   Remove the front wheels.
3   Remove the bolts and detach the undercover.
4   Remove the front stabilizer bar on models with power steering.
5   Disconnect the tie-rod ends from the steering knuckle assembly (Section 14).
6   Disconnect the pressure line on the power steering pump. Be prepared to catch the hydraulic fluid running out of the pump or tank and power steering gear. To insure that all the fluid has been drained from the system, hold the hydraulic line lower than the steering system and turn the steering wheel from lock-to-lock several times.
7   Remove the nut/bolt from the lower universal joint **(see illustration)**.
8   Loosen but do not remove the four mounting bolts on the steering gear **(see illustration)** until the shaft coming out of the steering gear can be disconnected from the steering universal joint.
9   Unscrew the pressure line clamp on the steering gear to gain access to the left lower mounting bolt.
10   Unscrew the return line on the steering gear.
11   Unscrew the four mounting bolts completely and remove the steering gear assembly.
12   Installation is the reverse of removal. Start the engine and bleed the steering system (Section 19). While the engine is running, check for leaks at the hose connections.
13   Have the front end alignment checked by a dealer service department or an alignment shop.

17   Steering wheel – removal and installation

Refer to illustrations 17.2, 17.4 and 17.5
1   Disconnect the negative cable at the battery.
Warning: *If equipped with air bag, have Porsche dealer disarm and remove prior to removing the steering wheel.*
2   Turn the steering wheel so that the wheels are pointing straight ahead and remove the steering wheel impact pad by pulling it off, starting at the lower outside corners to avoid damaging the horn electrical contacts **(see illustration)**.
3   Disconnect the horn wire.
4   Mark the steering shaft and wheel to ensure correct alignment during installation **(see illustration)**.
5   Remove the nut and spring washer from the steering column **(see illustration)**.
6   Remove the steering wheel.
7   Installation is the reverse of removal. Be sure to align the marks you made during removal.
8   Tighten the steering wheel nut to the specified torque.
9   Connect the battery cable and check the operation of the horn and turn signals. If there are any electrical malfunctions, refer to Chapter 12.

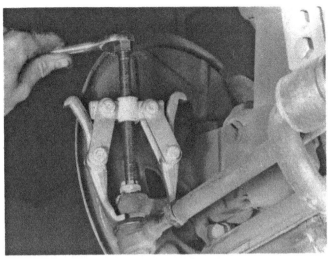

14.4 When separating the tie-rod end from the steering knuckle, loosen but do not remove the castellated nut

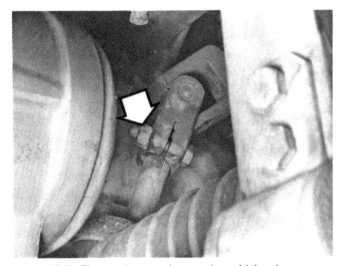

16.7 The steering gear lower universal joint clamp bolt/nut (arrow)

16.8 Once the steering gear is disconnected from the lower universal joint, the steering gear assembly may be released by removing the two clamps (arrows)

17.2 Lift the steering wheel pad, being careful not to damage the horn contacts

17.4 Note the paint marks on the wheel hub and shaft (arrows) to aid in correct steering wheel alignment during reassembly

17.5 Remove the steering wheel retaining nut

18.3   The steering column switch plate retaining screws
(arrows)

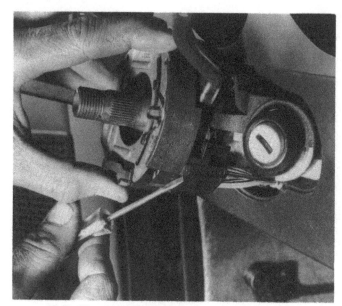

18.5   The electrical contacts on the steering column
switch assembly are easily damaged — exercise care when
prying off the connectors

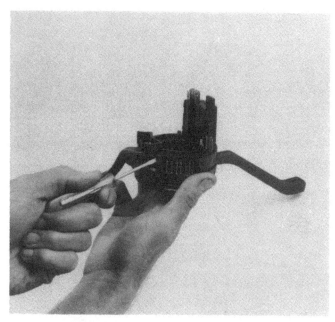

18.6   The two halves of the switch assembly are held
together by three small plastic tabs — carefully pry the
tabs free to separate the two assemblies

18   Steering column switches — removal and installation

Refer to illustrations 18.3, 18.5 and 18.6

1   Disconnect the negative cable at the battery. Place the cable out
of the way so it cannot accidentally come in contact with the negative
terminal of the battery.
2   Remove the steering wheel (Section 17).
3   Remove the switch plate retaining screws (see illustration).
4   Remove the two housing screws from underneath the switch housing. Separate the two housing halves and remove them from the
column.
5   Pry the electrical connectors from the back of the switch assembly
(see illustration) and pull the assembly off the column.

6   The two switch assemblies can be separated by prying the three
plastic locking tabs free (see illustration).
7   Installation is the reverse of removal.

19   Power steering system — bleeding

1   The power steering system must be bled whenever a line is disconnected. On later models, low fluid level can cause air to mix in with
the fluid resulting in a noisy pump as well as foaming of the fluid.
2   Open the hood and check the fluid level in the reservoir, adding the
specified fluid necessary to bring it up to the proper level (Chapter 1).
3   With the power steering pump and fluid at normal operating temperature, start the engine and slowly turn the steering wheel several
times from lock-to-lock. Check the fluid level, topping it up as necessary
until it remains steady and no more bubbles appear in the reservoir.
4   When starting the engine, the fluid level in the reservoir must not
rise by more than 10 mm (0.37-inch). If the fluid levels for a stopped
engine and a running engine deviate from each other by more than specified, there is too much air trapped in the hydraulic system. Repeat
the above procedure until the deviation is within specified limits.

20   Power steering pump — removal and installation

Removal

1   Remove the engine undercover.
2   Get a plastic pan or some rags to catch the hydraulic fluid, then
disconnect the pressure line from the bottom of the power steering
pump. The pump is located at the lower right front of the engine and
is driven by a toothed drivebelt. Do not reuse the power steering fluid.
3   Loosen the clamp on the intake hose at the top of the pump and
carefully disconnect the hose.
4   Disconnect the strut bolted to the bottom of the pump. On most
models, it is only necessary to disconnect the strut at the pump end.
5   Remove the two upper mounting bolts from the top right side of the
pump. This will release tension on the drivebelt so it can be removed.
6   Remove the long pivot/mounting bolt and detach the pump.

Installation

7   If the pulley was removed from the power steering pump, use a thread adhesive on the mounting bolts.

8   Installation is the reverse of the removal procedure. Check and adjust the drivebelt tension according to the procedure in Chapter 1.

9   Make sure the routing of the pressure line and intake hose is identical to the original installation. This will prevent friction and possible damage to these components.

10 Fill the power steering system with hydraulic fluid, bleed the system (Section 19) and check for leaks.

11 Test the function of the power steering by maneuvering the vehicle in a safe area before returning it to normal operation.

---

21   Steering angles and wheel alignment — general information

---

Proper wheel alignment is essential to proper steering and even tire wear. Symptoms of alignment problems are pulling of the steering to one side or the other and uneven tire wear.

If these symptoms are present, check for the following before having the alignment adjusted:

Loose steering gear mounting bolts
Damaged or worn steering gear mounts
Misadjusted steering gear
Worn or damaged wheel bearings
Bent steering tie-rods
Worn balljoints
Improper tire pressures
Mixing tires of different construction

Alignment faults in the rear suspension show up as uneven tire wear or uneven tracking of the rear wheels. This can be easily checked by driving the vehicle straight across a puddle of water onto a dry patch of pavement. If the rear wheels do not follow the front wheels exactly, the alignment should be adjusted.

Front or rear wheel alignment should be left to a dealer service department or an alignment shop.

Chapter 11 Body

Contents

Specifications

Torque specifications

	Ft-lbs	Nm
Front bumper impact bar-to-skid plate	15 to 25	20 to 34
Front bumper impact bar-to-front crossmember	15 to 25	20 to 34
Roof locator bracket adjusting bolts	15	20

---

1 General information

This Chapter confines itself to maintenance, repair and replacement procedures easily accomplished by the owner. It does not include procedures requiring the knowledge or training of a professional body and frame repair specialist.

The Porsche 944 Body design evolved out of the 924 model, which was based on research findings gained from the VW Experimental Safety Vehicle (ESV). "Crumple zones" are designed into the fore and aft portions of the body to absorb extreme impacts with minimum distortion to the passenger compartment. Legally mandated energy absorbing bumpers compliment this safety feature.

The factory paint is considered among the best finishes on any vehicle available at present and consists of a fade resistant, electromechanically bonded color coat and two clear coats of baked polyurethane enamel. Body repairs should always be finished with a urethane based enamel to match the fade rate of the original paint.

Many of the body components (sunroof, windows, hatch lock etc.) are electrically driven. Procedures addressing the electrical components of these body parts may be found in Chapter 12.

2 Body — general maintenance

1   The condition of your vehicle's body is very important, because it is an important component of the resale value of the vehicle. It is much more difficult to repair a neglected or damaged body than it is to repair mechanical components. And the condition of the hidden areas of the body (the fender wells, the frame, and the engine compartment, to name a few) — even though they don't require as much attention as the rest of the body — is just as important as the condition of more obvious exterior panels.
2   Once a year, or every 12,000 miles, it is a good idea to have the underside of the body and the frame steam cleaned. All traces of dirt and oil will be removed and the underside can then be inspected carefully for damaged brake lines, frayed electrical wiring, damaged cables, and other problems. The suspension components should be greased after completion of this job.
3   At the same time, clean the engine and the engine compartment using either a steam cleaner or a water soluble degreaser.

4   The body should be washed once a week (or when dirty). Wet the vehicle thoroughly to soften the dirt, then wash it down with a soft sponge and plenty of clean soapy water. If the surplus dirt is not washed off very carefully, it will eventually wear down the paint.
5   Spots of tar or asphalt coating thrown up from the road should be removed with a cloth soaked in solvent.
6   Once every six months, apply wax to the body.

3 Body damage — minor repair

See color photo sequence "Repair of minor scratches"

1   If the scratch is superficial and does not penetrate to the metal of the body, repair is very simple. Lightly rub the scratched area with a fine rubbing compound to remove loose paint and built up wax. Rinse the area with clean water.
2   Apply touch-up paint to the scratch, using a small brush. Continue to apply thin layers of paint until the surface of the paint in the scratch is level with the surrounding paint. Allow the new paint at least two weeks to harden, then blend it into the surrounding paint by rubbing with a very fine rubbing compound. Finally, apply a coat of wax to the scratch area.
3   If the scratch has penetrated the paint and exposed the metal of the body, causing the metal to rust, a different repair technique is required. Remove all loose rust from the bottom of the scratch with a pocket knife, then apply rust inhibiting paint to prevent the formation of rust in the future. Using a rubber or nylon applicator, coat the scratched area with glaze-type filler. If required, the filler can be mixed with thinner to provide a very thin paste, which is ideal for filling narrow scratches. Before the glaze filler in the scratch hardens, wrap a piece of smooth cotton cloth around the tip of a finger. Dip the cloth in thinner and then quickly wipe it along the surface of the scratch. This will ensure that the surface of the filler is slightly hollow. The scratch can now be painted over as described earlier in this section.

Repair of dents

4   When repairing dents, the first job is to pull the dent out until the affected area is as close as possible to its original shape. There is no point in trying to restore the original shape completely as the metal in the damaged area will have stretched on impact and cannot be restored to its original contours. It is better to bring the level of the dent

up to a point which is about 1/8-inch below the level of the surrounding metal. In cases where the dent is very shallow, it is not worth trying to pull it out at all.

5   If the back side of the dent is accessible, it can be hammered out gently from behind using a soft-face hammer. While doing this, hold a block of wood firmly against the opposite side of the metal to absorb the hammer blows and prevent the metal from being stretched.

6   If the dent is in a section of the body which has double layers, or some other factor makes it inaccessible from behind, a different technique is required. Drill several small holes through the metal inside the damaged area, particularly in the deeper sections. Screw long, self tapping screws into the holes just enough for them to get a good grip in the metal. Now the dent can be pulled out by pulling on the protruding heads of the screws with locking pliers.

7   The next stage of repair is the removal of paint from the damaged area and from an inch or so of the surrounding metal. This is easily done with a wire brush or sanding disk in a drill motor, although it can be done just as effectively by hand with sandpaper. To complete the preparation for filling, score the surface of the bare metal with a screwdriver or the tang of a file or drill small holes in the affected area. This will provide a good grip for the filler material. To complete the repair, see the Section on filling and painting.

Repair of rust holes or gashes

8   Remove all paint from the affected area and from an inch or so of the surrounding metal using a sanding disk or wire brush mounted in a drill motor. If these are not available, a few sheets of sandpaper will do the job just as effectively.

9   With the paint removed, you will be able to determine the severity of the corrosion and decide whether to replace the whole panel, if possible, or repair the affected area. New body panels are not as expensive as most people think and it is often quicker to install a new panel than to repair large areas of rust.

10   Remove all trim pieces from the affected area except those which will act as a guide to the original shape of the damaged body, such as headlight shells, etc. Using metal snips or a hacksaw blade, remove all loose metal and any other metal that is badly affected by rust. Hammer the edges of the hole inward to create a slight depression for the filler material.

11   Wire brush the affected area to remove the powdery rust from the surface of the metal. If the back of the rusted area is accessible, treat it with rust-inhibiting paint.

12   Before filling is done, block the hole in some way. This can be done with sheet metal riveted or screwed into place, or by stuffing the hole with wire mesh.

13   Once the hole is blocked off, the affected area can be filled and painted. See the following sub-section on filling and painting.

Filling and painting

14   Many types of body fillers are available, but generally speaking, body repair kits which contain filler paste and a tube of resin hardener are best for this type of repair work. A wide, flexible plastic or nylon applicator will be necessary for imparting a smooth and contoured finish to the surface of the filler material. Mix up a small amount of filler on a clean piece of wood or cardboard (use the hardener sparingly). Follow the manufacturer's instructions on the package, otherwise the filler will set incorrectly.

15   Using the applicator, apply the filler paste to the prepared area. Draw the applicator across the surface of the filler to achieve the desired contour and to level the filler surface. As soon as a contour that approximates the original one is achieved, stop working the paste. If you continue, the paste will begin to stick to the applicator. Continue to add thin layers of paste at 20-minute intervals until the level of the filler is just above the surrounding metal.

16   Once the filler has hardened, the excess can be removed with a body file. From then on, progressively finer grades of sandpaper should be used, starting with a 180-grit paper and finishing with 600-grit wet-or-dry paper. Always wrap the sandpaper around a flat rubber or wooden block, otherwise the surface of the filler will not be completely flat. During the sanding of the filler surface, the wet-or-dry paper should be periodically rinsed in water. This will ensure that a very smooth finish is produced in the final stage.

17   At this point, the repair area should be surrounded by a ring of bare metal, which in turn should be encircled by the finely feathered edge of good paint. Rinse the repair area with clean water until all of the dust produced by the sanding operation is gone.

18   Spray the entire area with a light coat of primer. This will reveal any imperfections in the surface of the filler. Repair the imperfections with fresh filler paste or glaze filler and once more smooth the surface with sandpaper. Repeat this spray-and-repair procedure until you are satisfied that the surface of the filler and the feathered edge of the paint are perfect. Rinse the area with clean water and allow it to dry completely.

19   The repair area is now ready for painting. Spray painting must be carried out in a warm, dry, windless and dust free atmosphere. These conditions can be created if you have access to a large indoor work area, but if you are forced to work in the open, you will have to pick the day very carefully. If you are working indoors, dousing the floor in the work area with water will help settle the dust which would otherwise be in the air. If the repair area is confined to one body panel, mask off the surrounding panels. This will help minimize the effects of a slight mismatch in paint color. Trim pieces such as chrome strips, door handles, etc., will also need to be masked off or removed. Use masking tape and several thicknesses of newspaper for the masking operations.

20   Before spraying, shake the paint can thoroughly, then spray a test area until the spray painting technique is mastered. Cover the repair area with a thick coat of primer. The thickness should be built up using several thin layers of primer rather than one thick one. Using 600-grit wet-or-dry sandpaper, rub down the surface of the primer until it is very smooth. While doing this, the work area should be thoroughly rinsed with water and the wet-or-dry sandpaper periodically rinsed as well. Allow the primer to dry before spraying additional coats.

21   Spray on the top coat, again building up the thickness by using several thin layers of paint. Begin spraying in the center of the repair area and then, using a circular motion, work out until the whole repair area and about two inches of the surrounding original paint is covered. Remove all masking material 10 to 15 minutes after spraying on the final coat of paint. Allow the new paint at least two weeks to harden, then use a very fine rubbing compound to blend the edges of the new paint into the existing paint. Finally, apply a coat of wax.

4 Body damage — major repair

1   In the event of a collision, it is essential that the frame and underbody be thoroughly checked and, if necessary, realigned by a body repair shop with the proper equipment.

2   Because each component contributes directly to the overall strength of the vehicle, proper materials and techniques must be used during body repair service. Again, because of the specialized materials, machinery, tools and techniques required for this kind of work, it should be performed by a professional paint and body shop.

3   Because some of the exterior panels are separate and replaceable units, they should be replaced rather than repaired if major damage occurs. Some of these components can be found in a wrecking yard that specializes in used vehicle components, often at considerable savings over the cost of new parts.

5 Upholstery and carpets — maintenance

1   Every three months remove the carpets or mats and clean the interior of the vehicle (more frequently if necessary). Vacuum the upholstery and carpets to remove loose dirt and dust.

2   Leather upholstery requires special care. Stains should be removed with warm water and a very mild soap solution. Use a clean, damp cloth to remove the soap, then wipe again with a dry cloth. Never use alcohol, gasoline, nail polish remover or thinner to clean leather upholstery.

3   After cleaning, regularly treat leather upholstery with a leather wax. *Never use car wax on leather upholstery.*

4   In areas where the interior of the vehicle is subject to hot, bright sunlight, it is advisable to cover leather seats with a sheet or other protective layer if the vehicle is to be left out for any length of time.

6 Vinyl trim — maintenance

Do not clean vinyl trim with detergents, caustic soap or petroleum-

These photos illustrate a method of repairing simple dents. They are intended to supplement *Body repair - minor damage* in this Chapter and should not be used as the sole instructions for body repair on these vehicles.

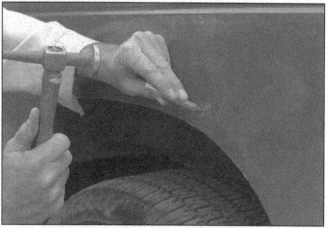

1 If you can't access the backside of the body panel to hammer out the dent, pull it out with a slide-hammer-type dent puller. In the deepest portion of the dent or along the crease line, drill or punch hole(s) at least one inch apart . . .

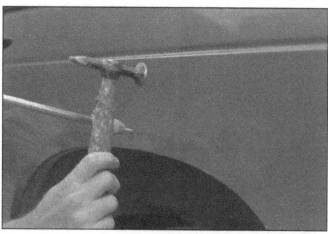

2 . . . then screw the slide-hammer into the hole and operate it. Tap with a hammer near the edge of the dent to help 'pop' the metal back to its original shape. When you're finished, the dent area should be close to its original contour and about 1/8-inch below the surface of the surrounding metal

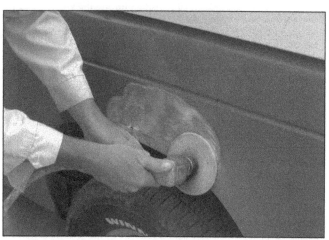

3 Using coarse-grit sandpaper, remove the paint down to the bare metal. Hand sanding works fine, but the disc sander shown here makes the job faster. Use finer (about 320-grit) sandpaper to feather-edge the paint at least one inch around the dent area

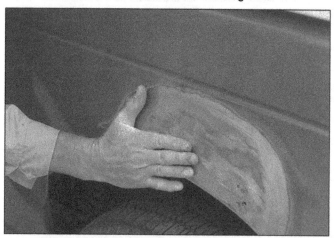

4 When the paint is removed, touch will probably be more helpful than sight for telling if the metal is straight. Hammer down the high spots or raise the low spots as necessary. Clean the repair area with wax/silicone remover

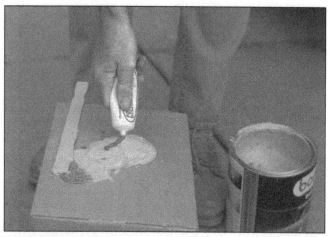

5 Following label instructions, mix up a batch of plastic filler and hardener. The ratio of filler to hardener is critical, and, if you mix it incorrectly, it will either not cure properly or cure too quickly (you won't have time to file and sand it into shape)

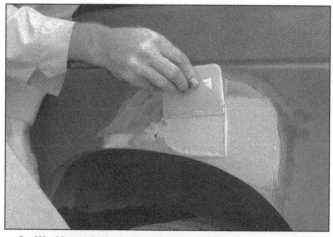

6 Working quickly so the filler doesn't harden, use a plastic applicator to press the body filler firmly into the metal, assuring it bonds completely. Work the filler until it matches the original contour and is slightly above the surrounding metal

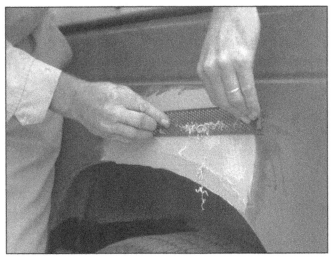

7   Let the filler harden until you can just dent it with your fingernail. Use a body file or Surform tool (shown here) to rough-shape the filler

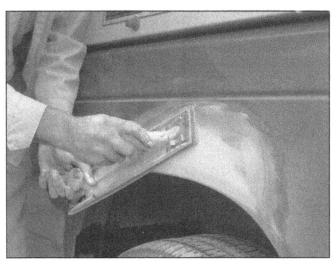

8   Use coarse-grit sandpaper and a sanding board or block to work the filler down until it's smooth and even. Work down to finer grits of sandpaper - always using a board or block - ending up with 360 or 400 grit

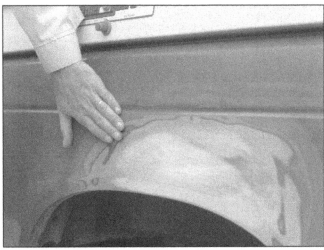

9   You shouldn't be able to feel any ridge at the transition from the filler to the bare metal or from the bare metal to the old paint. As soon as the repair is flat and uniform, remove the dust and mask off the adjacent panels or trim pieces

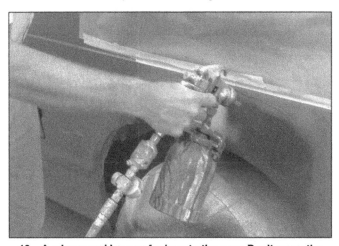

10   Apply several layers of primer to the area. Don't spray the primer on too heavy, so it sags or runs, and make sure each coat is dry before you spray on the next one. A professional-type spray gun is being used here, but aerosol spray primer is available inexpensively from auto parts stores

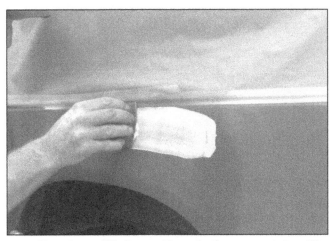

11   The primer will help reveal imperfections or scratches. Fill these with glazing compound. Follow the label instructions and sand it with 360 or 400-grit sandpaper until it's smooth. Repeat the glazing, sanding and respraying until the primer reveals a perfectly smooth surface

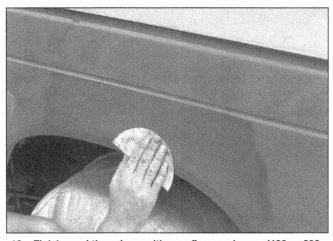

12   Finish sand the primer with very fine sandpaper (400 or 600-grit) to remove the primer overspray. Clean the area with water and allow it to dry. Use a tack rag to remove any dust, then apply the finish coat. Don't attempt to rub out or wax the repair area until the paint has dried completely (at least two weeks)

based cleaners. Plain soap and water is all that is required, with a soft brush to remove dirt that may be ingrained. Wash the vinyl as frequently as the rest of the vehicle.

After cleaning, application of a high quality rubber and vinyl protectant will help prevent oxidation and cracking. This protectant can also be applied to weatherstripping, vacuum lines, tires and rubber hoses, which often fail as a result of chemical degradation.

7   Exterior mouldings – removal and installation

1   Some 944 models come equipped with optional door side guards. In the event that these become damaged or deteriorated, they must be removed with extreme care to avoid damaging the paint.
2   Begin removal of the door side guard at the edge of the door nearest the door lock. Drill through the plastic rivet with a 4 mm (0.160/5/32-inch) diameter drill.
3   Peel off the side guard carefully, using a sharp knife to help separate the moulding from the door.
4   Once the moulding is detached, remove the remaining adhesive stripping with a sharp knife, being careful to avoid paint damage. Remove as much of the adhesive as possible with the knife, then dissolve the rest of it with alcohol.
5   Wash and dry the vehicle. Do not install the new side guard mouldings in cold or moist weather. Ambient temperature should be at least 68 °F (20 °C).
6   Fill the adhesive groove in the back of the moulding with adhesive compound (Porsche part No. AMV 176 000 05) so that the bead is approximately 1 mm higher than the contour of the moulding. Install the

strips no longer than 10 minutes from the time the adhesive is applied.
7   Peel off the backing from the adhesive strip on the new moulding and lightly press it into place. Do not touch the adhesive surface. When the moulding is correctly positioned, press the moulding firmly with a roller.
8   Secure the end of the moulding strip with a new plastic rivet or sheet metal screw.

8   Stone guard strips — replacement

Refer to illustration 8.3

Note: *Ambient air temperature should be no less than 68 °F (20 °C) when replacing the stone guard strips.*

1   Porsche utilizes transparent plastic stone protection sheets on vulnerable areas around the perimeter of the wheel wells. If they are damaged or deteriorated, they can be replaced.
2   An industrial heat gun is preferred for heating the stone guards for removal. A conventional hair blow drier may also be used. The latter method is truer to the glamorous spirit of the 944, though it takes more time.
3   Heat the entire adhesive surface and peel away the strips **(see illustration)**.
4   Remove the dirt from the adhesive surfaces on the fender and clean them with alcohol to remove all adhesive.
5   Moisten the adhesive surface of the fender with a 50/50 mixture of alcohol and water. Pull off the paper backing from the new strips

8.3   The stone guard adhesive should be heated to a gum-like consistency before peeling the plastic sheet away with a putty knife

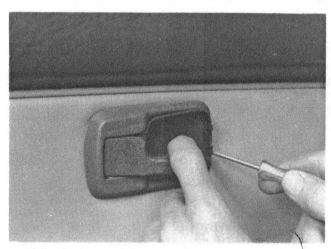

10.2   The door latch finger guard is removed by prying the tabs loose

10.3   Underneath the finger guard is the door latch housing, retained by a single screw

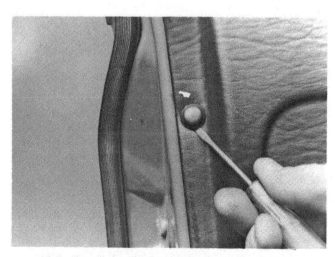

10.4   Pry off the plastic caps to access the door trim panel screws

and apply them to the fenders in the appropriate position.

6   Working from the center of the strip outward, massage out moisture and air bubbles with a plastic spatula or similar tool. Make sure no air bubbles remain trapped.

7   Obstinate bubbles may be removed by puncturing them with a pin and pressing out the air.

8   Remove excess adhesive from the edges of the strips with alcohol after the adhesive has begun to dry.

9   Hinges and locks — maintenance

Once every 3000 miles, or every three months, the hinges, locks and latch assemblies on the doors, hood and rear hatch should be given a few drops of light oil or lock lubricant. Powdered graphite lubricant is preferred for lock cylinders, where petroleum-based lubricants may gather dirt and other abrasives. The door latch strikers should also be lubricated with a thin coat of grease to reduce wear and ensure free movement.

10   Door trim panel — removal and installation

Refer to illustrations 10.2, 10.3, 10.4, 10.5, 10.6, 10.7, 10.8a and 10.8b

1   Unscrew and remove the door lock button.

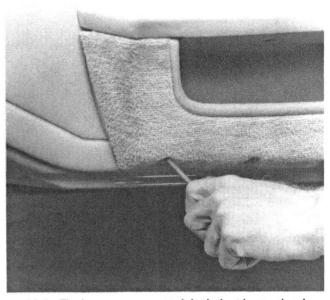

10.5   The bottom screws attach both the trim panel and the door glove box to the door frame

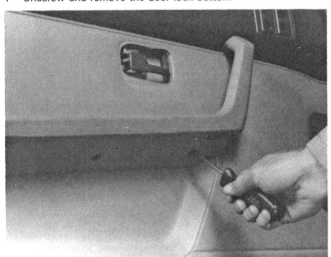

10.6   It is not necessary to disconnect the upper armrest mount after removing the three lower mounting screws

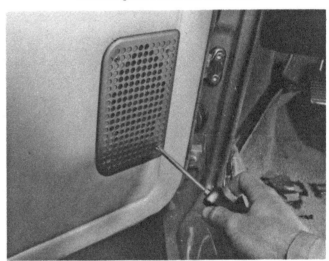

10.7   Some models require the removal of the speaker grille before the trim panel can be removed

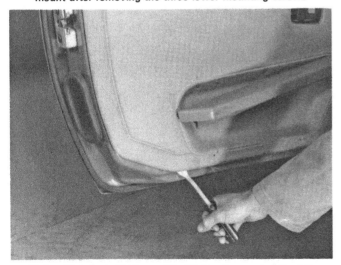

10.8a   Use the widest possible blade when prying the trim panel from the door frame to prevent distortion of the panel

10.8b   The connectors for the electric mirror and window controls must be removed before the trim panel can be detached

2   Carefully pry the door latch finger guard loose and remove it from the door **(see illustration)**.
3   Remove the door latch housing retaining screw and pull the housing from the door **(see illustration)**.
4   Separate the soft plastic caps from the trim panel retaining screws and remove the screws **(see illustration)**.
5   Remove the retaining screws from the door panel glove box and remove the glove box **(see illustration)**.
6   Remove the armrest retaining screws and let the armrest hang by the upper mount **(see illustration)**.
7   Remove the speaker grille retaining screws and the speaker grille **(see illustration)**.
8   Beginning at the front of the door, carefully pry the door clips from the door frame **(see illustration)**. The trim panel may now be lifted up and away from the door to allow disconnection of the power window and mirror controls **(see illustration)** and the speaker wires.
9   Installation is the reverse of removal. Note that the trim panel support must hook over the door frame when dropping the trim panel into place. Test the window and mirror controls before installing the trim panel.
10 Make certain the clips are aligned, then strike each one firmly with the heel of your hand to seat them.

---

11   Door handle — removal and installation

Refer to illustrations 11.2, 11.3 and 11.4
1   Remove the door trim panel (Section 10).

2   With a small screwdriver, spread the plastic clip at the end of the door lock linkage **(see illustration)** and disconnect the linkage.
3   Remove the door handle retaining screw from the door jamb **(see illustration)**.
4   Pull the door handle away from the door enough to separate the lock linkage **(see illustration)** and remove the door handle from the vehicle.
5   Installation is the reverse of the removal procedure. Lubricate all the moving parts of the door handle and linkage assembly before installing it.

---

12   Door lock cylinder — removal and installation

Refer to illustrations 12.2, 12.3, 12.4 and 12.5
1   Insert the key into the lock cylinder to prevent the locking plates from flying all over the place during removal. Keep it in place throughout the rest of this procedure.
2   Remove the retaining screw from the end of the lock cylinder **(see illustration)**.
3   Pry off the two plates the screw was holding in place **(see illustration)**. Note that they are spring-loaded and must be installed in exactly the same way.
4   Remove the return spring, noting how it is installed for reassembly **(see illustration)**.
5   Remove the lock cylinder from the door handle, making sure the key is fully inserted in the lock **(see illustration)**.
6   Reverse the above procedure for installation. You can take the key

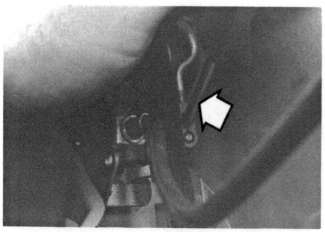

11.2   Disconnect the door lock linkage by spreading the plastic connector with a small screwdriver (arrow)

11.3   This screw (arrow) must be removed to separate the door handle from the door

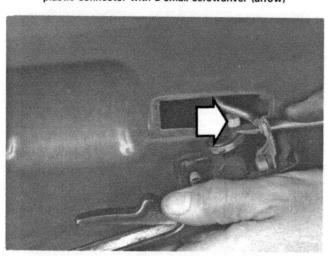

11.4   The final step in removing the door handle is to disconnect the white plastic piece from the linkage (arrow)

12.2   Removing the screw from the end of the lock cylinder will allow the spring-loaded components to be detached for lock cylinder removal — note that the key is in the lock throughout the procedure

out after the handle is completely assembled. Lubricate the lock cylinder with powdered graphite or other lock cylinder lubricant.

13 Door striker — adjustment

1   The door striker is mounted in the body pillar, threaded into a floating cage. To adjust the striker in an inboard-outboard or up-down direction, loosen the Allen head striker bolts and shift the striker as required (see illustration 11.3).

2   It may be helpful to mark the location of the striker on the pillar before loosening it, to use as a frame of reference.

3   Close the door until the striker just touches the latch and note what adjustment is required to align the center of the striker bar with the center of the opening in the latch. Move the striker, tighten the bolts and open and close the door a few times.

4   When you are done, apply a light coat of general purpose grease to the striker bar where it contacts the latch.

14 Door mounted mirrors — removal and installation

Refer to illustrations 14.2 and 14.3

1   Remove the internal mirror components from the mirror housing (Chapter 12).

2   Push the ends of the wires out of the plastic connector (see illustration). Make a diagram of the connector so you know where the wires go when you put them back in.

3   Use an Allen wrench to remove the mirror housing from the door panel (see illustration).

4   Installation is the reverse of removal.

12.3   Note the way the spring loaded components are assembled before taking them off the door handle

12.4   Remove the crank return spring . . .

12.5   . . . and pull the key and lock cylinder assembly from the door handle

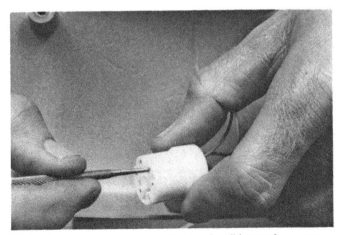

14.2   Before the mirror housing can be slid over the motor drive wire harness, the pin connectors must be pushed out carefully with a suitable tool

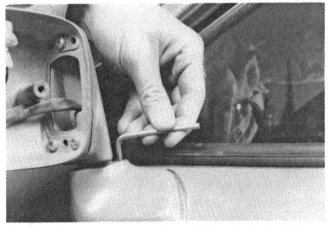

14.3   The mirror housing is attached to the body with a single Allen bolt accessible by turning the mirror clockwise as far as possible

15 Bumpers — removal and installation

Refer to illustrations 15.2 and 15.3

Front

1   Two bolts hold the front bumper to the vehicle. They are accessible from underneath, above the fog lamps.
2   Remove the two bolts with a socket wrench **(see illustration)**.
3   Extend the bumper just enough to disconnect the lights on each

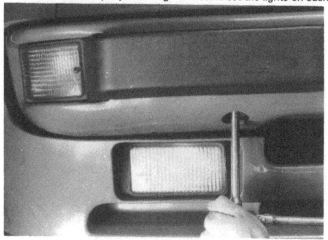

15.2   The front bumper mounting bolts are removed through the access holes

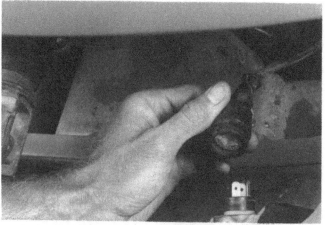

15.3   Disconnect the light wire harness connectors before pulling the bumper clear of the vehicle

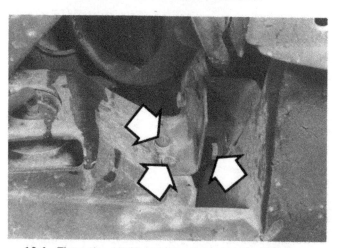

16.4   These three bolts must be removed from each lower corner of the radiator bracket (arrows)

end **(see illustration)**.
4   Pull the bumper away from the vehicle.
5   Installation is the reverse of removal. Clean the contacts on the electrical connectors and replace the rubber boots if they need it.

Rear

6   Remove the four nuts holding the plastic bumper end pieces to the main body. From this point, the rear bumper comes off just like the front one.
7   Remove the two rear bumper mount bolts and lift the bumper away from the vehicle.
8   Installation is the reverse of removal.

16 Front spoiler — removal and installation

Refer to illustrations 16.3, 16.4 and 16.5

1   Raise the front of the vehicle and secure it on jackstands.
2   Unplug the electrical connectors from the side marker and driving lights.
3   Remove the mounting bolts from the outboard portions of the spoiler **(see illustration)**.
4   Remove the three bolts on each side attaching the spoiler to the radiator brackets **(see illustration)**.
5   Remove the final three bolts attaching the spoiler to the radiator lower crossmember **(see illustration)**.
6   Remove the spoiler from the vehicle, checking to make sure all remaining electrical components have been disconnected.
7   To install the spoiler, reverse the removal procedure. Clean the electrical contacts on each connector and replace any worn-out rubber boots.

16.3   A typical spoiler has three bolts (arrows) and two electrical connectors (arrows) on each side of the vehicle

16.5   Three bolts attach the spoiler to the crossmember (arrows)

17 Hood — removal, installation and adjustment

Refer to illustrations 17.5 and 17.14

Removal

1 Open the hood and disconnect the negative cable from the battery.
2 Place a fender cover in the opening between the hood and nose to protect the paint.
3 Disconnect the wire harness for the underhood service light.
4 Disconnect the windshield washer fluid tube.
5 Paint marks or scribe a line around the hood hinge bolts and the perimeter of the hood hinges **(see illustration)**.
6 Remove the hood support struts by disconnecting them at the top. Remove the retaining clip with a small screwdriver and then pull out the pin. To prevent loss of the pin and retaining clip, refit them to the strut and let the strut rest in the engine compartment.
7 Remove the hood hinge bolts and, with the aid of an assistant, lift off the hood. Note the number of shims used on each hinge, if any.

Installation

8 With the aid of an assistant, put the hood in place, install any shims which were removed earlier and loosely install the hinge-to-hood bolts.
9 Align the hinges with the scribe marks made earlier and tighten the hinge bolts.
10 Install the hood support struts.
11 Reconnect the service light and windshield washer fluid feed tube.
12 Connect the ground cable to the battery.

Adjustment

13 The hood is adjusted by loosening the hood hinge bolts slightly, moving the hood, then retightening the bolts.
14 Adjust the hood latch by loosening the bolts which attach the latch to the hood **(see illustration)**, repositioning the latch, then tightening the bolts.

18 Rear hatch — removal, installation and adjustment

Refer to illustrations 18.4, 18.5 and 18.10

Removal

1 Open the rear hatch.
2 Unplug the heater element wire.
3 Grasp the body of the strut and unscrew it until the strut separates from the ball pivot socket. After removing both struts, set the hatch down easily to prevent the latch from closing.
4 Remove the two plastic hinge covers from the rear edge of the roof **(see illustration)**.
5 Remove the two hatch mounting bolts from each hinge **(see illustration)**.
6 With the help of an assistant, lift the hatch from the vehicle.

Installation and adjustment

7 Installation is the reverse of removal. Have your assistant hold the hatch in place while loosely installing the hinge bolts. Press the hatch into place until it locks and move the hatch around until it is perfectly aligned. Without unlocking the latch, tighten the bolts.

17.5   Mark the position of the hood hinges and bolts before loosening the bolts

17.14   The hood latch retaining bolts, when loosened slightly, allow for alignment (extension of the male latch fitting is made with a screwdriver at the nose cone [arrow])

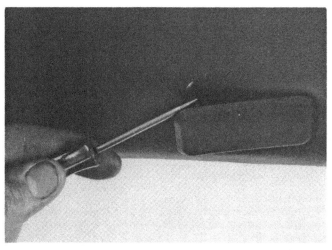

18.4   The rear hatch hinge covers can be pried loose from the headliner . . .

18.5   . . . to access the hinge mounting bolts

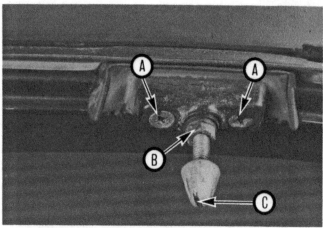

**18.10   Loosen the latch retaining screws (A) to make
lateral adjustments; loosen the jam nut (B) and adjust
the latch extension by turning the nose cone (C)
with a screwdriver**

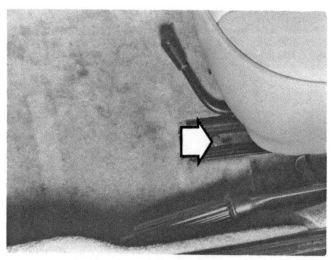

19.2   Driver's seat rail left front retaining bolt (arrow)

8   Reconnect the hatch struts.
9   Unlock the latch and carefully open the hatch to check for binding
and insufficient clearance, especially on the top edge where the hinges
attach. If binding occurs, loosen the hinge bolts and readjust the hatch.
10 If clearances around the perimeter of the hatch are correct, but
the latches are not aligning properly, slightly loosen the two screws
on each male latch component and relocate them **(see illustration)**.
11 The extension of the male latch fitting may be altered by loosening
the jam nut and screwing the fitting in or out, then tightening the jam
nut. This adjustment controls how snug the hatch fits against the rubber
seal.

19   Seats — removal and installation

Refer to illustrations 19.1 and 19.2

1   Slide the seat forward and remove the rear bolts **(see illustration)**.
2   Slide the seat back and remove the front bolts **(see illustration)**.
3   If the vehicle is equipped with power seats, unplug the wire harness
and remove the seat from the vehicle.
4   Check the seat rails for wear and replace them if necessary.
5   Lubricate the seat rails with a light coat of general purpose grease
before installation.
6   Place the seat on the rail carriers. If the vehicle has power seats,
connect the wiring harness at this time.

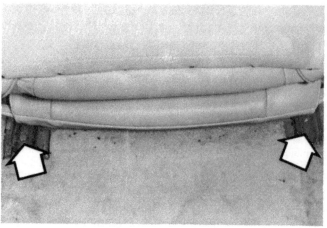

19.1   Seat rail rear bolts (arrows)

7   Align the front of the seat with the holes. Install and lightly tighten
the front bolts.
8   Slide the seat forward and lock it into position, then install the rear
bolts and tighten them slightly.
9   Slide the seat back to the lock position and completely tighten the
forward bolts.
10 Lock the seat in the forward position and tighten the rear bolts.

20   Window glass — removal and installation

Replacing window glass is a procedure requiring specialized tools
and knowledge, so we recommend that glass be replaced by a dealer
service department or an automotive glass shop.

21   Headlight assembly — removal and installation

1   Disconnect the negative cable at the battery.
2   Raise the headlights by turning the knob on top of the motor
(Chapter 12, Section 5). If a rubber shroud is installed, it will have to
be removed to access the knob. Check the operation of the bellcranks,
bearings and mounts. If binding is noted in the bellcrank assembly when
the raising knob is operated, check the condition of the bearings and
the adjustment of the cranks. If the bellcrank assembly is binding, have
a dealer service department correct the alignment of the components.
This requires a special jig.
3   If the headlights do not raise when the switch is turned on, there
are two checks which should be performed before contemplating replacement of the motor.
4   Check the plug for good contact and clean terminals. Clean any
rusted or corroded terminals with emery paper or fine wet/dry sandpaper.
5   Check the condition of the relay mounted on the side of the motor.
An auto electric shop should be consulted when checking the relay.
If the relay is in good condition, replace the motor.
6   Disconnect the plug from the motor and remove the relay.
7   Remove the nut at the end of the operating shaft, then remove
the three bolts which hold the motor to the raising assembly.
8   Before installing the new motor, connect the power plug and the
relay, turn the headlight switch to the Off position and touch the
negative lead to the negative post of the battery. This will move the
motor to its off/lowered position.
9   Install the motor and secure all bolts and nuts previously removed.
10 Connect the negative cable to the battery and raise the headlights.
If they do not raise fully, adjust the operating lever by pressing it onto
different splines on the motor shaft.
11 Check the headlight adjustment. When the headlights are raised,
the rubber bumper should rest on the sheet metal so that the headlights
are firmly held in place. If the headlight beams are in the approximate
height position, adjust the rubber bumpers by screwing them in or out
as required.
12 Adjust the headlight beams as described in Chapter 12.

Chapter 12   Chassis electrical system

Contents

1   General information

The electrical system is a 12-volt, negative ground type. Power for the lights and all electrical accessories is supplied by a lead/acid battery which is charged by the alternator.

This Chapter covers repair and service procedures for the various electrical components not associated with the engine. Information on the battery, alternator and starter motor can be found in Chapter 5.

It should be noted that whenever portions of the electrical system are serviced, the negative battery cable should be disconnected at the battery to prevent electrical shorts and/or fires.

2   Electrical troubleshooting — general information

A typical electrical circuit consists of an electrical component, any switches, relays, motors, etc. related to that component and the wiring and connectors that connect the component to both the battery and the chassis. To aid in locating a problem in any electrical circuit, wiring diagrams are included at the end of this book.

Before tackling any troublesome electrical circuit, first study the appropriate diagrams to get a complete understanding of what makes up that individual circuit. Trouble spots, for instance, can often be narrowed down by noting if other components related to that circuit are operating properly or not. If several components or circuits fail at one time, chances are the problem lies in the fuse or ground connection, as several circuits are often routed through the same fuse and ground connections.

Electrical problems often stem from simple causes, such as loose or corroded connections, a blown fuse or a melted fusible link. Always visually inspect the condition of the fuse, wires and connections in a problem circuit before troubleshooting it.

If testing instruments are going to be utilized, use the diagrams to plan ahead of time where you will make the necessary connections in order to accurately pinpoint the trouble spot.

The basic tools needed for electrical troubleshooting include a circuit tester or voltmeter (a 12-volt bulb with a set of test leads can also be used), a continuity tester, which includes a bulb, battery and set of test leads, and a jumper wire, preferably with a circuit breaker incorporated, which can be used to bypass electrical components.

Voltage checks should be performed if a circuit is not functioning properly.

Connect one lead of a circuit tester to either the negative battery terminal or a known good ground. Connect the other lead to a connector in the circuit being tested, preferably nearest to the battery or fuse. If the bulb of the tester lights up, voltage is present, which means that the part of the circuit between the connector and the battery is problem free. Continue checking the rest of the circuit in the same fashion. When you reach a point at which no voltage is present, the problem lies between that point and the last test point with voltage. Most of the time the problem can be traced to a loose connection. **Note:** *Keep in mind that some circuits receive voltage only when the ignition key is in the Accessory or Run position.*

One method of finding shorts in a circuit is to remove the fuse and connect a test light or voltmeter in its place to the fuse terminals. There should be no voltage present in the circuit. Move the wiring harness from side-to-side while watching the test light. If the bulb goes on, there is a short to ground somewhere in that area, probably where the insulation has rubbed through. The same test can be performed on each component in the circuit, even a switch.

Perform a ground test to check whether a component is properly grounded. Disconnect the battery and connect one lead of a self-powered test light, known as a "continuity tester," to a known good ground. Connect the other lead to the wire or ground connection being tested. If the bulb goes on, the ground is good. If the bulb does not go on, the ground is not good.

A continuity check determines if there are any breaks in a circuit — if it is conducting electricity properly. With the circuit off (no power in the circuit), a self-powered continuity tester can be used to check the circuit. Connect the test leads to both ends of the circuit (or to the "power" end and a good ground), and if the test light comes on the circuit is passing current properly. If the light doesn't come on, there is a break somewhere in the circuit. The same procedure can be used to test a switch, by connecting the continuity tester to the power in and power out sides of the switch. With the switch turned on, the test light should come on.

When diagnosing for possible open circuits, it is often difficult to locate them by sight because oxidation or terminal misalignment are hidden by the connectors. Merely wiggling a connector on a sensor or in the wiring harness may correct the open circuit condition. Remember this if an open circuit is indicated when troubleshooting a circuit. Intermittent problems may also be caused by oxidized or loose connections.

Electrical troubleshooting is simple if you keep in mind that all electrical circuits are basically electricity running from the battery, through the wires, switches, relays, fuses and fusible links to each electrical component (light bulb, motor, etc.) and back to ground, from which it is passed back to the battery. Any electrical problem is an interruption in the flow of electricity to and from the battery.

---

3 Fuses — general information

The electrical circuits of the vehicle are protected by a combination of fuses. Pre-1984 models incorporate all fuses into a single fuse block. Models manufactured since 1984 utilize a Central Electric Board, which combines the lower amperage fuses with the relays for circuits which were formerly in various remote locations. Vehicles with the Central Electric Board also incorporate an auxiliary fuse box for higher amperage components such as motors, blowers and the rear window defogger. The Central Electric Board is located above the firewall in the engine compartment just below the driver's side windshield wiper.

Each fuse protects one or more circuits. The protected circuit is identified by a number on the fuse panel face right next to each fuse. Miniaturized fuses are employed in both the fuse block and the Central Electric Board. Later models have a relay equipped with a test connection on top so that the fuses can be checked. If a fuse is inserted in the test connection and the fuse is operational, an LED light will come on. See the printed layout on the underside of the fuse panel lid for correct identification of this relay and for identification of circuits serviced by a particular fuse or relay.

If an electrical component fails, always check the fuse first. A blown fuse, which is nothing more than a broken element, is easily identified. Visually inspect the element for evidence of damage. If a continuity check is called for, use the test connection on the relay mentioned above.

Be sure to replace blown fuses with the correct type. Fuses of different ratings are physically interchangeable, but only fuses of the proper rating should be used. Replacing a fuse with one of a higher or lower value than specified is not recommended. Each electrical circuit needs a specific amount of protection. The amperage value of each fuse is molded into the fuse body. **Caution:** *At no time should a fuse be bypassed with pieces of metal. Serious damage to the electrical system could result.*

If the replacement fuse immediately fails, do not replace it again until the cause of the problem is isolated and corrected. In most cases, this will be a short circuit in the wiring caused by a broken or deteriorated wire.

---

4 Exterior bulb replacement

Refer to illustrations 4.3, 4.7 and 4.9

Front turn signals

1   Remove the two screws attaching the turn signal lens to the bumper. Press the bulb into the holder, turn it counterclockwise and remove it from the unit. The bulb may then be replaced.
2   When replacing the lens, make sure the gasket is in good condition. Tightening the lens screws too tightly may crack the lens, so be careful.

Rear light cluster

3   Pull back the interior carpet in the cargo area. The bulb holder may be removed by unscrewing the white knurled knob and pulling the housing off **(see illustration)**.
4   Do not remove the lens from the body unless replacing it. If it is being replaced, be sure the rubber gasket is in good condition and that the seal is watertight before replacing the bulb holder and carpet.

License plate light

5   Remove the lens retaining screws and detach the lens. The bulb may be replaced by pressing it in and turning it counterclockwise. Check the condition of the rubber grommet before reassembling the unit.

Front side markers

6   Raise the front of the vehicle and support it on jackstands.
7   Unscrew the two plastic knurled knobs and pull off the rear portion of the side marker assembly **(see illustration)**. The bulb can be removed

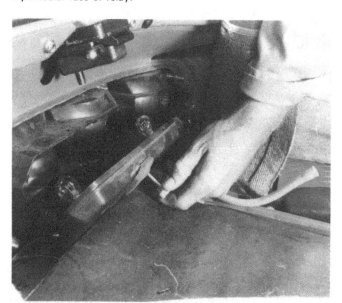

4.3   The rear light cluster bulb housing comes off after unscrewing the knurled knob

4.7   The front side marker light is held in place by two knurled knobs (arrows)

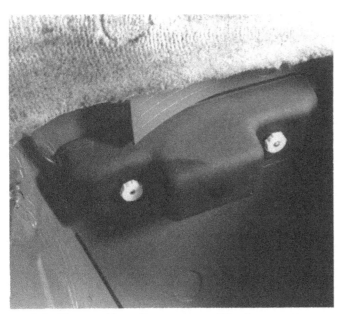

4.9 The rear side markers are accessible after removing the carpet from the rear corners of the cargo area

5.3 The emergency headlight raising knob is located atop the drive motor, under the protective rubber boot

without removing the lens from the vehicle body. Make sure the seal is watertight when replacing the backing plate.

Rear side markers

8 Pull away the interior carpet from the corner of the cargo area.
9 Unscrew the two knurled knobs and pull off the rear of the side marker assembly **(see illustration)**.
10 The bulb and bracket may be remove by pressing in on the plastic lever at the front of the side marker assembly.

---

5 Headlight doors — emergency manual operation

Refer to illustration 5.3

1 In the case of a failure in the headlight raising system, the headlights may be raised manually.

2 Open the hood.
3 Remove the protective rubber boot from the top of the headlight raising system motor **(see illustration)**.
4 Turn the knurled knob to raise or lower the headlight assemblies. If undue resistance is encountered, lubricate the retractor linkage.

---

6 Headlights — removal and installation

Refer to illustrations 6.3, 6.4 and 6.5

1 Raise the headlights.
2 Disconnect the negative cable from the battery.
3 Remove the screws from the headlight trim **(see illustration)** and detach it.
4 Remove the screws from the sealed beam carrier frame **(see illustration)**. Pull the headlight directly toward you to remove it from the frame.

6.3 The first step in removing or adjusting the headlight is to remove the trim

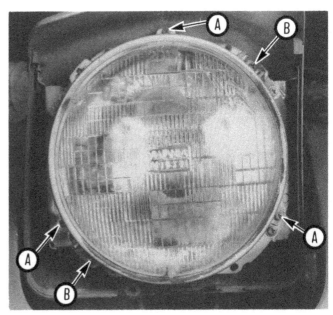

6.4 The headlight sealed beam unit may be removed from the frame by removing the retaining screws (A); two spring-loaded screws are used to adjust the headlight (B)

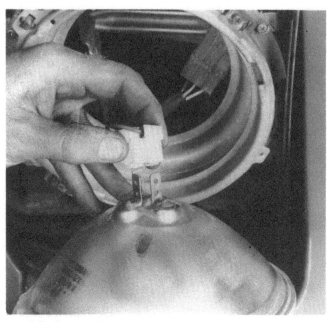

6.5   Remove the wire harness connector from the back of the sealed beam unit

5   Disconnect the wire harness **(see illustration)**.
6   Installation is the reverse of the removal procedure. Adjust the headlights (Section 7) after replacement.

7   Headlights — adjustment

Note: *It is important that the headlights be aimed correctly. If adjusted incorrectly they could blind the driver of an oncoming vehicle and cause an accident or seriously reduce your ability to see the road. The headlights should be checked for proper aim every 12 months and any time*

a new sealed beam headlight is installed or front end body work is performed. It should be emphasized that the following procedure is only an interim step which will provide temporary adjustment until the headlights can be adjusted by a properly equipped shop.

1   Headlights have two spring loaded adjusting screws, one on the top, controlling up-and-down movement and one on the side, controlling left-and-right movement **(see illustration 6.4)**.
2   There are several methods of adjusting the headlights. The simplest method requires a blank wall 25 feet in front of the vehicle and a level floor.
3   Park the vehicle 25 feet from the wall.
4   Position masking tape vertically on the wall in reference to the vehicle centerline and the centerlines of both headlights.
5   Position a horizontal tape line in reference to the centerline of each of the headlights. It may be easier to position the tape on the wall with the vehicle parked only a few inches away.
6   Adjustment should be made with the vehicle sitting level, the gas tank half-full and no unusually heavy load in the vehicle.
7   Starting with the low beam adjustment, position the high intensity zone so it is two inches below the horizontal line and two inches to the right of the headlight vertical line. Adjustment is made by turning the top adjusting screw *clockwise* to raise the beam and *counterclockwise* to lower the beam. The adjusting screw on the side should be used in the same manner to move the beam left or right.
8   With the high beams on, the high intensity zone should be vertically centered with the exact center just below the horizontal line. **Note:** *It may not be possible to position the headlight aim exactly for both high and low beams. If a compromise must be used, keep in mind that the low beams are the most used and have the greatest effect on driver safety.*
9   Have the headlights adjusted by a dealer service department at the earliest opportunity.

8   Door mounted mirrors — component replacement

Refer to illustrations 8.1, 8.2, 8.3 and 8.4

Pre-1985 models

1   Very carefully separate the mirror glass from the drive mechanism plate with a screwdriver **(see illustration)**.
2   Disconnect the heated mirror electrical wires **(see illustration)**.

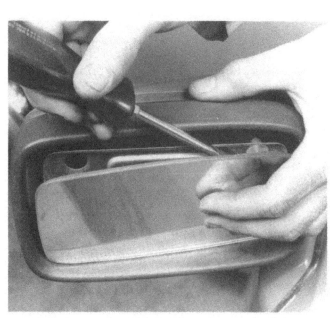

8.1   Exteme care must be taken when prying off the mirror glass on older models

8.2   Removing the mirror glass reveals the heated mirror grid and the two wires which must be detached before removing the glass

8.3   Three retaining bolts hold the mirror drive mechanism to the housing

8.4   Separate the two connector halves and the drive mechanism can be removed from the housing

3   Remove the three drive mechanism retaining bolts and pull the drive mechanism away from the mirror housing **(see illustration)**.
4   Disconnect the main electrical connector **(see illustration)** and remove the drive mechanism from the vehicle.
5   To remove the mirror housing from the body, see the procedure in Chapter 11.
6   Installation is the reverse of removal.

1985 and later models

7   The mirror glass on these models is mounted differently than on earlier models and resembles a bayonet fastener.
8   To separate the mirror glass from the base, insert a small screwdriver into the hole at the base of the mirror housing and turn the retaining ring on the mirror glass counterclockwise. The ring has teeth into which one can insert the screwdriver. The glass on both left and right mirrors is identical.
9   Separate the glass from the mirror housing enough to disconnect the electrical plug (heated mirrors only), then remove the glass from the mirror housing.
10   The rest of the procedure is the same as described in the procedure above. Behind the mirror glass is the drive mechanism. This can be separated from the mirror housing by removing the three mirror drive retaining bolts and disconnecting the electrical plug from the drive.
11   Installation is the reverse of the removal procedure.

9   Horn — removal and installation

Refer to illustration 9.2

Removal

1   Disconnect the negative cable from the battery.
2   Pull off the electrical connectors at each horn **(see illustration)**.
3   Remove the horn mounting bolt from the horn bracket and detach the horn assembly.

Installation

4   Installation is the reverse of removal. Do not mount the horn body so that it contacts any part of the vehicle other than the horn mounting bracket as this may cause vibration or rattling.

10   Radio antenna — removal and installation

Refer to illustrations 10.4, 10.7 and 10.8
Note: *Later models are equipped with a windshield-integrated antenna which cannot be removed. Should a problem arise with this type of antenna, repair it using the same general method as described for repairing the rear window defogger (Section 13).*

1   Raise the vehicle and support it on jackstands.
2   Remove the left front wheel.
3   Remove the two retaining bolts from the emissions canister and swing it out of the way.

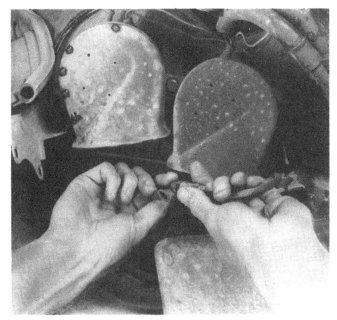

9.2   Disconnect the electrical connector before removing the horn assembly bracket bolt

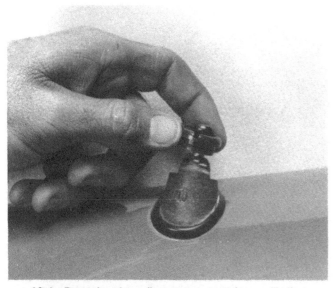

10.4   Removing the radio antenna escutcheon will allow the antenna assembly to be pulled through the bottom of the fender

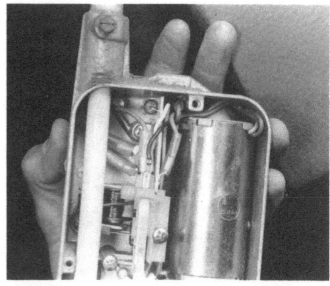

10.7   After removing the cover plate from the power antenna drive unit, the electrical leads can be disconnected

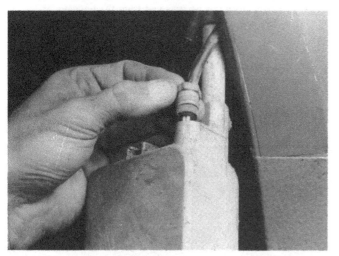

10.8   The final step is to unscrew the power lead locknut and pull the wires from the drive unit

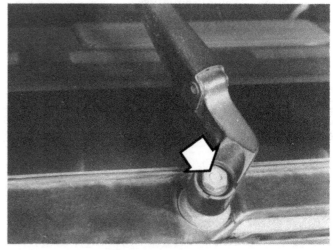

11.2   The windshield wiper arm is held in place by a nut (arrow)

4   Unscrew the retaining ring (escutcheon) from the top side of the antenna assembly (see illustration) and the plastic cone from underneath.
5   Remove the slotted bolt from the antenna lower mounting bracket.
6   Pull the power antenna assembly away from the fender enough to unscrew the antenna lead.
7   Remove the four screws from the cover plate on the power antenna unit and detach the cover plate. Disconnect the three electrical leads from the connectors inside the unit (see illustration), noting their location for reassembly.
8   Unscrew the locknut where the power cable enters the antenna unit and pull the wire and rubber locking grommet from the top of the unit (see illustration). Remove the antenna unit from the vehicle.
9   Installation is the reverse of removal

---

11   Windshield wiper arm — removal and installation

Refer to illustration 11.2
Note: *See the procedure in Chapter 1 for information on wiper blade inspection and replacement.*

Removal
1   Turn the ignition key to the On position and activate the windshield wiper switch. Turn off the ignition switch when the arms are about halfway through their arcs.
2   After prying off the plastic cap at the base of the wiper arm, use a socket to remove the wiper arm retaining nut (see illustration).
3   Mark the relationship of the wiper arm assembly to the spindle for correct installation.
4   Pry the wiper arm up on alternate sides to "walk" the wiper arm off the spindle.
5   Detach the wiper arm.

Installation
6   Position the wiper arm assembly on the spindle, aligning the marks you made prior to removal. Lightly tighten the retaining nut.
7   Turn the ignition key to the On position, cycle the wipers once and check the gap between the tip of the wiper arm and the windshield post at the end of the arm's sweep.

12.3   The windshield wiper motor is located under a
plastic cover which is retained by a single screw

12.5   The windshield wiper linkage is fastened to the
motor with a single nut

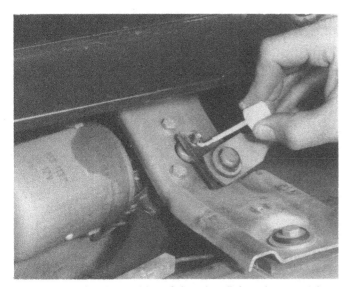

12.6   Marking the position of the wiper linkage is essential
before disconnecting it completely

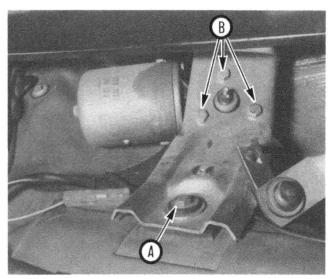

12.8   The motor can be removed after first removing the
bracket retaining bolt (A) and the three motor mounting
bolts (B)

12   Windshield wiper motor — removal and installation

Refer to illustrations 12.3, 12.5, 12.6 and 12.8

Pre-1985 models

1   Disconnect the negative cable from the battery.
2   Remove the windshield wiper arm (Section 11).
3   Remove the retaining screw from the plastic shield above the wiper motor and detach the shield **(see illustration)**.
4   Unplug the electrical connector from the wiper motor terminals.
5   Remove the linkage retaining nut **(see illustration)** but do not remove the linkage at this time.
6   Mark the position of the wiper linkage on the drive shaft **(see illustration)**.
7   Remove the wiper linkage from the drive shaft.
8   Remove the wiper motor mounting bracket retaining bolt and the three motor bolts **(see illustration)**. The bracket may now be raised and the motor removed.
9   If the motor is defective, replace it. There are no serviceable parts for this item. Take the old motor with you to the parts store for identification.

10   Inspect the linkage for wear or corrosion. Lubricate all moving parts with general purpose grease. Linkage parts are replaceable.
11   Installation is the reverse of removal.

1985 and later models

12   Removing and installing the wiper motor is the same as above with the following differences.
13   The wiper motor is located next to the Central Electric Board, below the driver's side windshield wiper arm.
14   After removing the black plastic shroud, the wiper motor can be separated from the wiper carrier frame assembly by removing the three attaching screws and disconnecting the electrical plug.

13   Rear window defogger — check and repair

Refer to illustrations 13.5 and 13.11

1   The rear window is equipped with a number of horizontal elements baked into the glass surface during the glass forming operation.
2   Small breaks in the element can be successfully repaired without

ZONES OF BULB BRILLIANCE

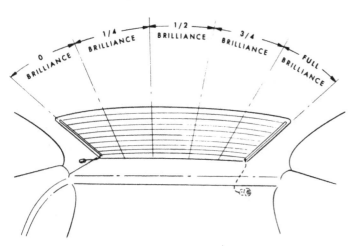

13.5   The brilliance of the test light should vary as shown when the rear window defogger is functioning normally

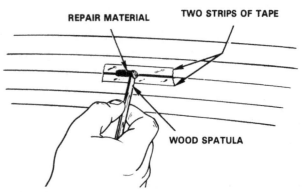

13.11   To repair a broken element, apply a strip of tape to both sides of the line, then apply the mixture of hardener and silver plastic with a small wooden spatula

removing the rear window.
3   To test the grids for proper operation, start the engine and turn on the system.
4   Ground one lead of a test light and carefully touch the other lead to each element line.
5   The brilliance of the test light should increase as the lead is moved across the element **(see illustration)**. If the test light glows brightly at both ends of the lines, check for a loose ground wire. All of the lines should be checked in at least two places.
6   To repair a break in a line, it is recommended that a repair kit specifically for this purpose be purchased. Included in the repair kit will be a decal, a container of silver plastic and hardener, a mixing stick and instructions.
7   To repair a break, first turn off the system and allow it to de-energize for a few minutes.
8   Lightly buff the element area with fine steel wool, then clean it thoroughly with alcohol.
9   Use the decal supplied in the repair kit or apply strips of electrician's tape above and below the area to be repaired. The space between the pieces of tape should be the same width as the existing lines. This can be checked from outside the vehicle. Press the tape tightly against the glass to prevent seepage.
10   Mix the hardener and silver plastic thoroughly.
11   Using the wood spatula, apply the silver plastic mixture between the pieces of tape, overlapping the undamaged area slightly on either end **(see illustration)**.
12   Carefully remove the decal or tape and apply a constant stream of hot air directly to the repaired area. A heat gun set at 500 to 700 degrees Fahrenheit is recommended. Hold the gun one inch from the glass for two minutes.
13   If the new element appears off color, tincture of iodine can be used to clean the repair and bring it back to the proper color. This mixture should not remain on the repair for more than 30 seconds.
14   Although the defogger is now fully operational, the repaired area should not be disturbed for at least 24 hours.

14   Speedometer speed sensor — replacement

1   Disconnect the cable from the negative terminal of the battery.
2   Raise the vehicle and support it securely on jackstands.
3   The speed sensor is located on the left side of the transmission, next to the driveaxle flange.
4   Disconnect the electrical connector from the speed sensor.
5   Use an open-end wrench to unscrew the sensor from the transmission.
6   Installation is the reverse of removal.

15   Electronic cruise control — general information

Electronic cruise control is an optional system which will maintain a driver-selected speed under normal driving conditions.
Because of obvious safety considerations, the cruise control system should be serviced only by trained professionals. There are no owner serviceable components in this system.

16   Power windows — general information

Because of the difficulty involved in gaining access to the power window components and because of the special tools required, servicing of the power windows should be done by a dealer service department.

17   Headlight switch — replacement

1   The headlight switch is located in the lower left corner of the dashboard, to the left of the instrument cluster.
2   Disconnect the cable from the negative terminal of the battery.
3   Pry around the perimeter of the switch with a flat tool, being careful not to damage the dash.
4   When the back of the switch is accessible, separate the switch from the connector.
5   Have the old switch tested by a dealer if you suspect that it's faulty, and take the old switch with you for identification if you are replacing it.
6   Installation is the reverse of removal.

18   Turn signal/headlight dimmer switch — replacement

Refer to illustrations 18.4a and 18.4b
1   Disconnect the negative cable at the battery.
2   Remove the steering wheel (Chapter 10).
3   Remove the switch assembly from the steering column (Chapter 10).
4   Remove the turn signal/headlight dimmer switch from the assembly by prying loose the three plastic locking tabs **(see illustration)** and separating the switch assembly halves **(see illustration)**.
5   If there is some chance that you have misdiagnosed the switch as the problem, take it to an auto electric shop or dealer service department for testing before buying a new one.
6   Installation is the reverse of the removal procedure.

19   Windshield wiper switch — replacement

1   Disconnect the negative cable at the battery.
2   Remove the steering wheel (Chapter 10).

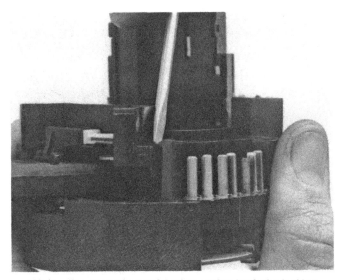

18.4a Be extremely careful when prying back the locking tabs on the switch assembly, as the tabs are brittle and easily broken

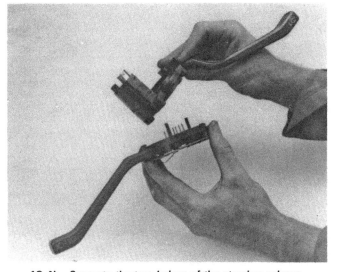

18.4b Separate the two halves of the steering column switch assembly after releasing the locking tabs

3   Remove the switch assembly from the steering column (Chapter 10).
4   Separate the windshield wiper switch from the assembly by prying the two halves apart at the three plastic locking tabs (see illustrations 18.4a and 18.4b).
5   Take the switch to an auto electric shop or dealer service department for testing if there is some chance it has been misidentified as the problem.
6   Installation is the reverse of the removal procedure.

20   Ignition switch assembly — removal and installation

Refer to illustrations 20.4a, 20.4b and 20.4c

1   Disconnect the cable from the negative terminal of the battery.
2   Remove the steering wheel (Chapter 10).
3   Remove the turn signal and windshield wiper switches as an assembly as shown in Chapter 10.
4   Follow the accompanying illustration sequence to complete the removal procedure (see illustrations).
5   Installation is the reverse of the removal procedure. Be sure the switch is in the On position before installing it on the steering column.

20.4a The first step in removing the ignition switch assembly is prying off the plastic sleeve

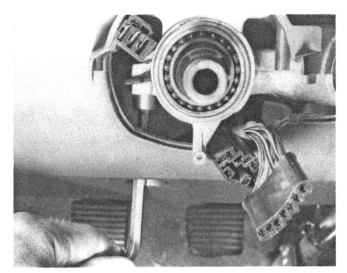

20.4b Remove the Allen head retaining bolt through the hole under the dash

20.4c The ignition switch unit may now be removed from the steering column shaft

21   Instrument cluster — removal and installation

Refer to illustration 21.2

Note: *Because of its solid state design, the instrument cluster assembly is not repairable by nor can it be disassembled by the home mechanic. It is also extremely expensive to replace. If a malfunction occurs in the instrument cluster assembly, it is recommended that it be taken to a dealer service department for repairs.*

1   Remove the steering wheel (Chapter 10).
2   Remove the two instrument cluster retaining screws **(see illustration)**.
3   Pull the cluster out of the dash far enough to disconnect the speedometer cable (earlier models only; models with electronic speedometer signal generator may skip this step).
4   Pull the cluster away from the dash. Note the arrangement of the electrical connectors.
5   Disconnect the electrical connectors and remove the instrument cluster from the vehicle. If testing is necessary, it is recommended that the unit be taken to a dealer service department.
6   Installation is the reverse of the removal procedure.

22   Power door locks — general information

The power door lock system consists of the Central Locking System Control Unit clipped to the lower end of the steering column, a microswitch at each door lock cylinder and actuation elements for each microswitch.

Before taking the switch or control unit out of the vehicle, check them with your test light. If you're certain a component is bad, remove it, take it to a dealer and get a new one. A dealer service department can test the circuits at the control unit for a reasonable cost.

The Central Locking System Control Unit is easily removed for testing by manually pulling the unit from near the base of the column. Remove

21.2   The instrument cluster is held in place by two screws in the outer instrument pods (arrows)

the bracket from the unit if you are replacing it since the new one probably won't come with one.

The microswitch at the lock cylinder is more likely to fail than the control unit. It is accessible after removing the door interior trim panel (Chapter 11). Disconnect the electrical plug, remove the outside door handle (Chapter 11) and then carefully press the microswitch off the handle with a screwdriver.

Installation is the reverse of removal.

DIAGRAM KEY

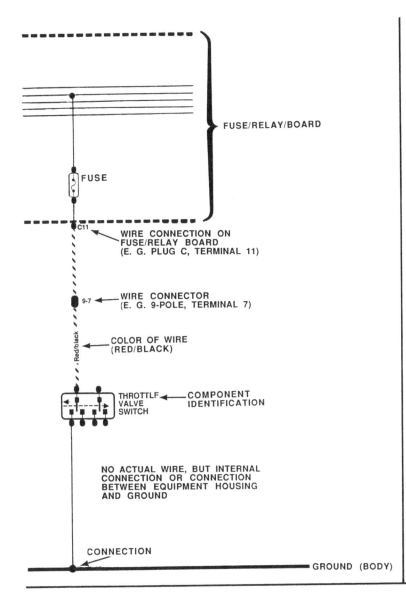

FUSE/RELAY/BOARD

FUSE

WIRE CONNECTION ON
FUSE/RELAY BOARD
(E. G. PLUG C, TERMINAL 11)

WIRE CONNECTOR
(E. G. 9-POLE, TERMINAL 7)

COLOR OF WIRE
(RED/BLACK)

THROTTLE
VALVE
SWITCH

COMPONENT
IDENTIFICATION

NO ACTUAL WIRE, BUT INTERNAL
CONNECTION OR CONNECTION
BETWEEN EQUIPMENT HOUSING
AND GROUND

CONNECTION

GROUND (BODY)

WIRE CONNECTORS

ONE-POLE
1 - A-NEAR FUSE/RELAY BOARD
    B-IN ENGINE COMPARTMENT
    C-NEAR BRAKE BOOSTER
    D-IN INSTRUMENT PANEL
    E-ON REAR LID
    H-NEAR DOOR SWITCH, RIGHT
    I - NEAR RADIO

TWO-POLE
2 - A-NEAR FRESH AIR BLOWER
    B-IN INSTRUMENT PANEL
    C-NEAR LEFT SEAT
    D-NEAR RIGHT REAR LIGHTS
    E-NEAR LEFT REAR LIGHTS
    F-NEAR FUSE/RELAY BOARD
    G-NEAR FUSE/RELAY BOARD
    H-NEAR LEFT HEADLIGHTS
    I - NEAR RIGHT HEADLIGHTS
    J-CENTER CONSOLE LIFT
    K-IN ENGINE COMPARTMENT
    L-NEAR FOG LIGHT, LEFT
    M-NEAR FOG LIGHT, RIGHT
    N-IN ENGINE COMPARTMENT

THREE-POLE
3 - B-NEAR HEADLIGHT, LEFT
    C-NEAR HEADLIGHT, RIGHT

FOUR-POLE
4 - NEAR FUSE/RELAY BOARD

SIX-POLE
6 - A-NEAR FUSE/RELAY BOARD
    B-IN CENTER CONSOLE

SEVEN-POLE
7 - B-NEAR RIGHT REAR LIGHTS
    C-NEAR LEFT REAR LIGHTS

NINE-POLE
9 - IN ENGINE COMPARTMENT

GROUND CONNECTION LOCATIONS

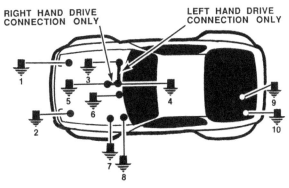

RIGHT HAND DRIVE
CONNECTION ONLY

LEFT HAND DRIVE
CONNECTION ONLY

FOR 1985-86 MODELS ONLY

0035-H

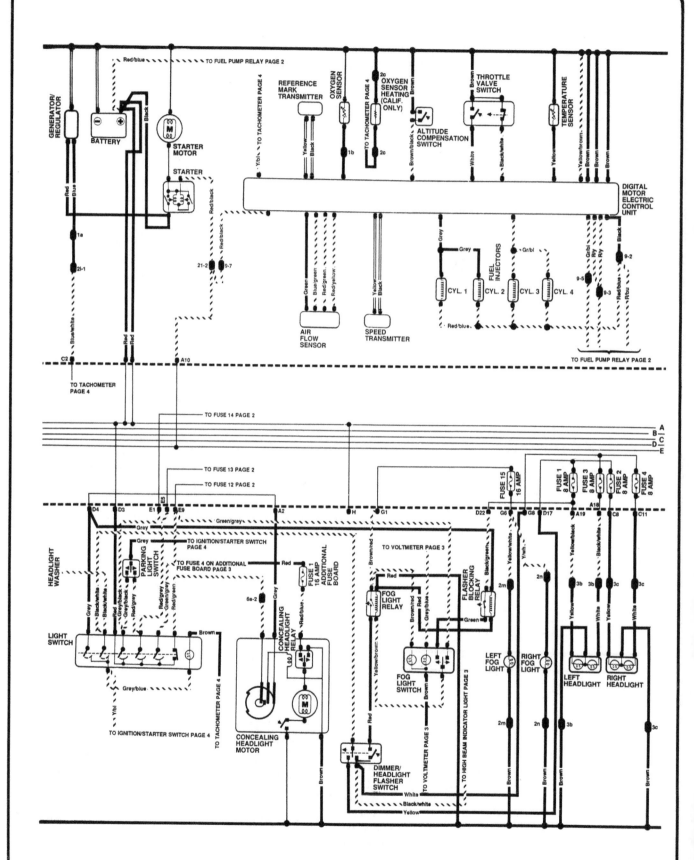

1983 models: Alternator, starter, oxygen sensor, throttle valve switch, temperature sensor, Digital Motor Electronic (DME) control unit, air flow sensor, fuel injectors, light switch, headlight motor, fog lights, headlights and dimmer switch
(Page 1 of 5)

0036-H

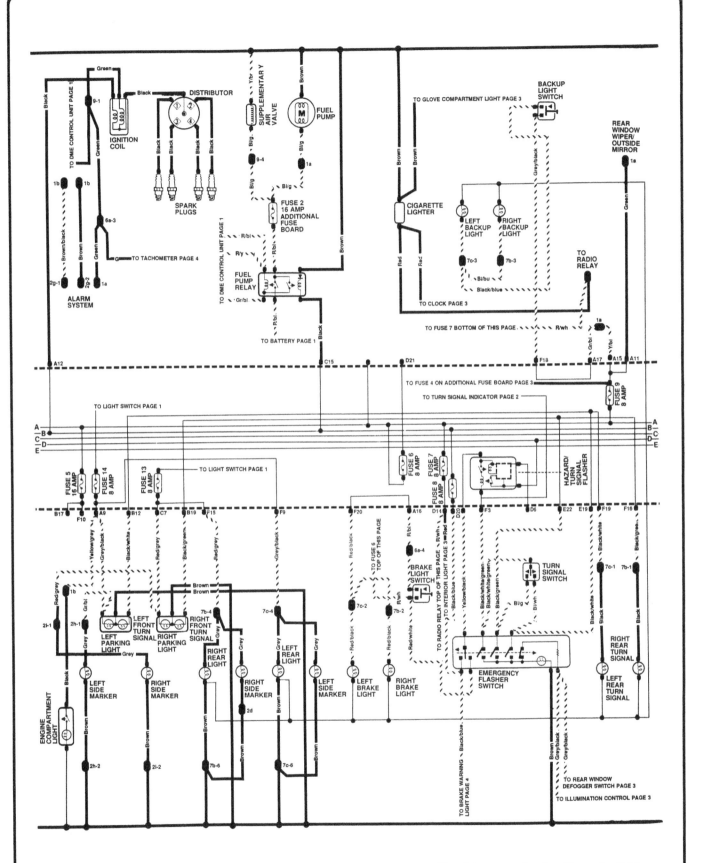

1983 models: Ignition system, fuel pump. cigarette lighter, back-up lights, parking lights, turn signals, side marker lights, flashers, turn signal switch and brake light switch (Page 2 of 5)

0037-H

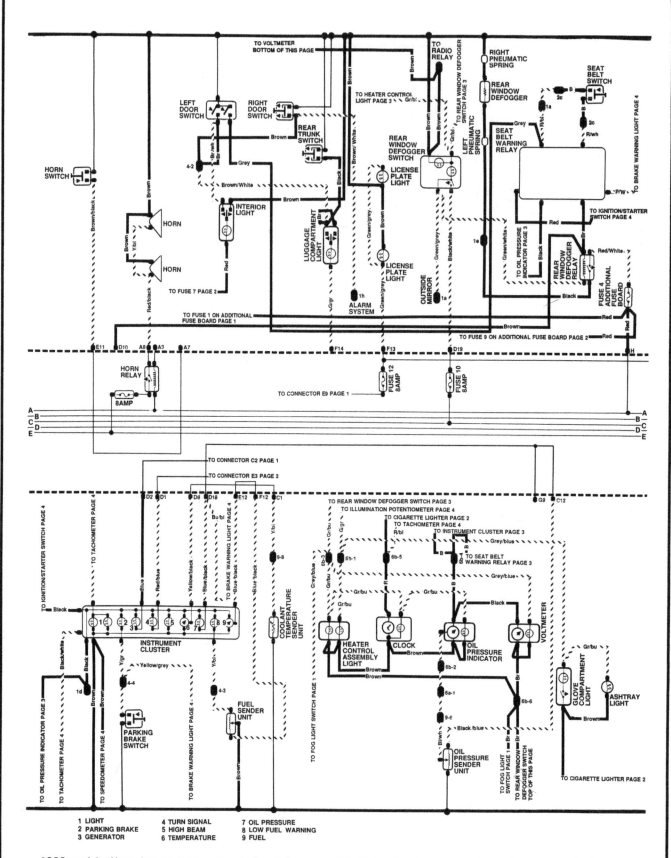

1 LIGHT   4 TURN SIGNAL   7 OIL PRESSURE
2 PARKING BRAKE 5 HIGH BEAM   8 LOW FUEL WARNING
3 GENERATOR   6 TEMPERATURE   9 FUEL

1983 models: Horn, interior lights, rear window defogger, seat belt warning, instrument cluster, heater control, clock and oil pressure sender (Page 3 of 5)

0038-H

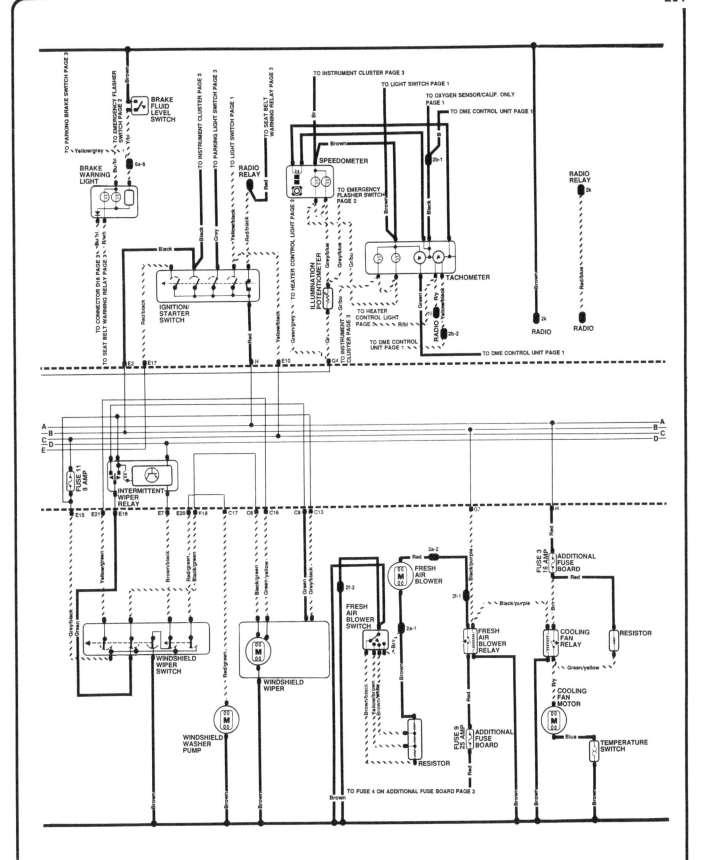

1983 models: Brake warning light, ignition switch, speedometer, tachometer, wiper relay, wiper switch, windshield wiper, blower switch, blower, temperature switch and cooling fan (Page 4 of 5)

0039-4

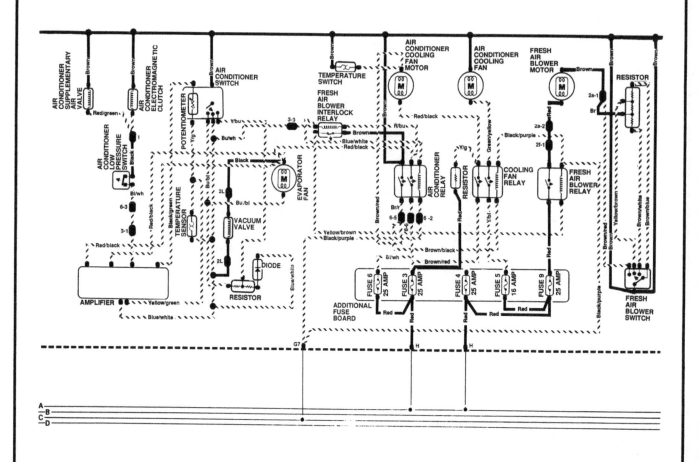

1983 models: Air conditioning (Page 5 of 5)

0040-H

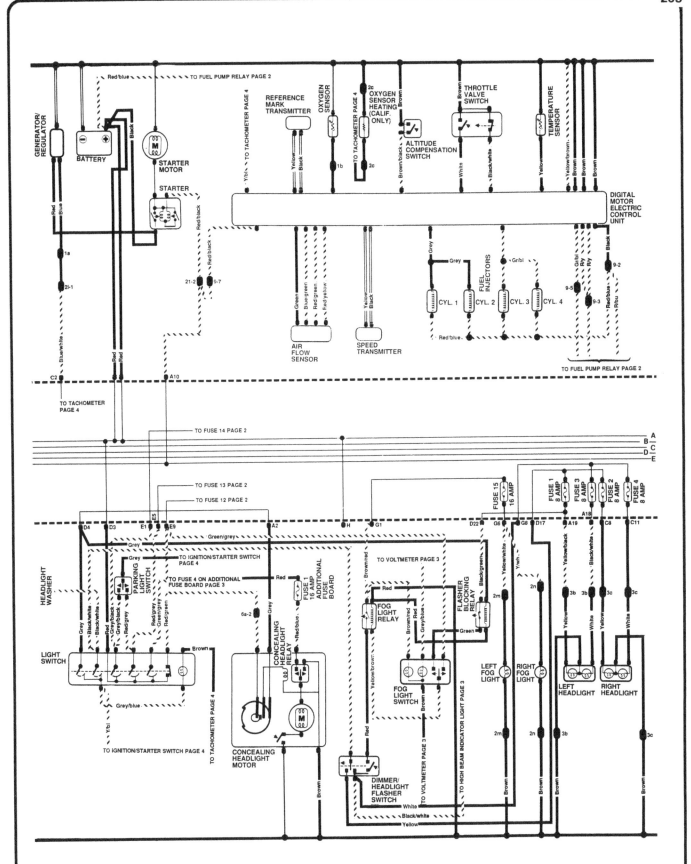

1984 models: Starter, alternator, oxygen sensor, throttle valve switch, temperature sensor, Digital Motor Electronic (DME) control unit, air flow sensor, fuel injectors, light switch, headlight motor, fog lights, dimmer switch and headlights
(Page 1 of 5)

0041-H

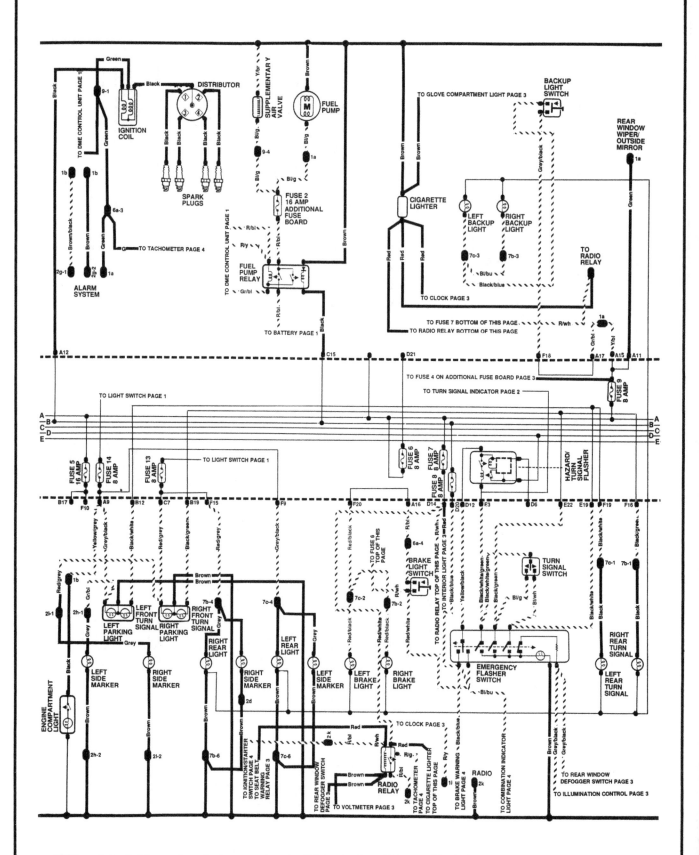

1984 models: Ignition system, fuel pump, cigarette lighter, back-up lights, parking lights, turn signals, side marker lights, brake lights, flashers, brake light switch and turn signal switch (Page 2 of 5)

0042-H

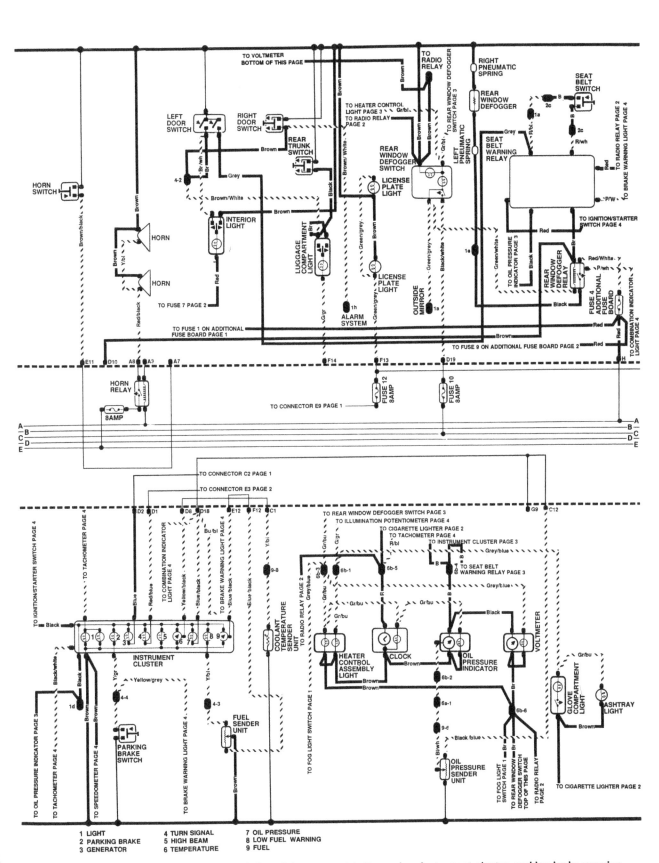

1984 models: Horn, interior lights, rear window defogger, seat belt warning, instrument cluster, parking brake warning, fuel sender, coolant temperature sender, heater control, clock and oil pressure sender (Page 3 of 5)

0043-H

206

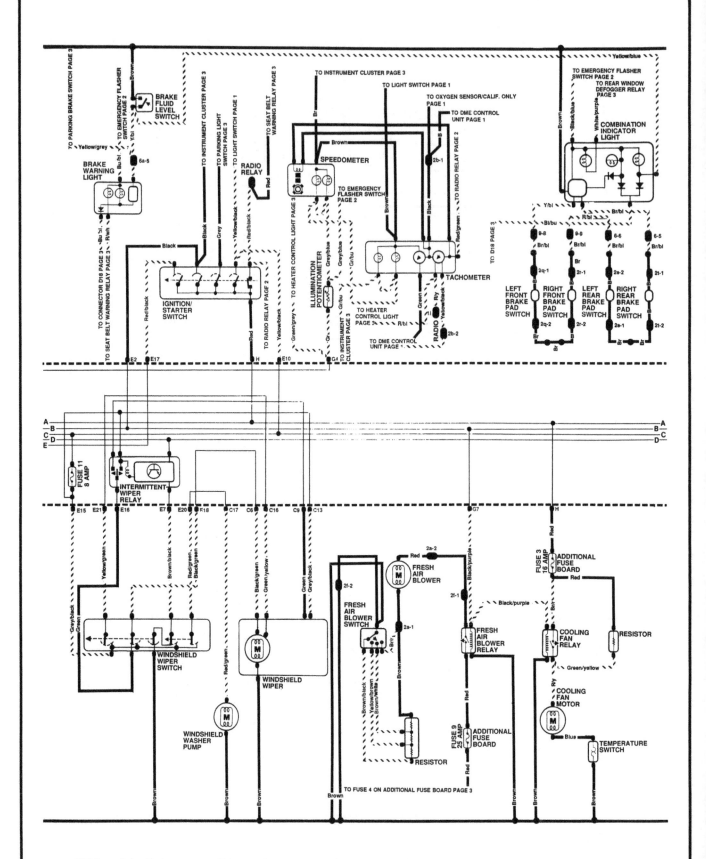

1984 models: Brake warning light, ignition switch, speedometer, tachometer, brake pad wear sensors, combination indicator light, wiper relay, wiper switch, windshield wiper, washer pump, blower, temperature switch and cooling fan
(Page 4 of 5)

0044-H

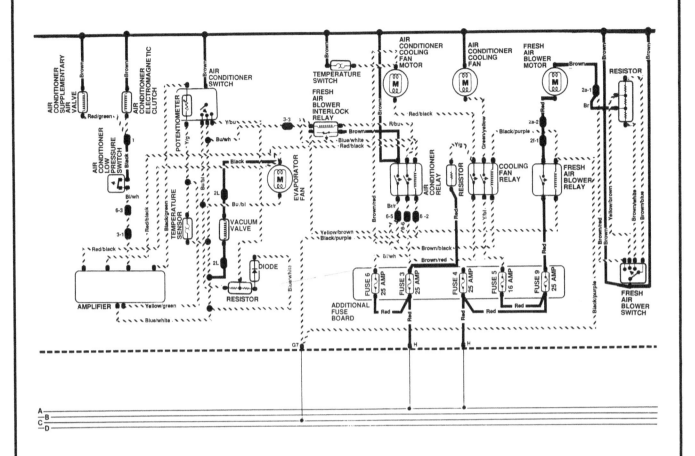

1984 models: Air conditioning (Page 5 of 5)

0045-H

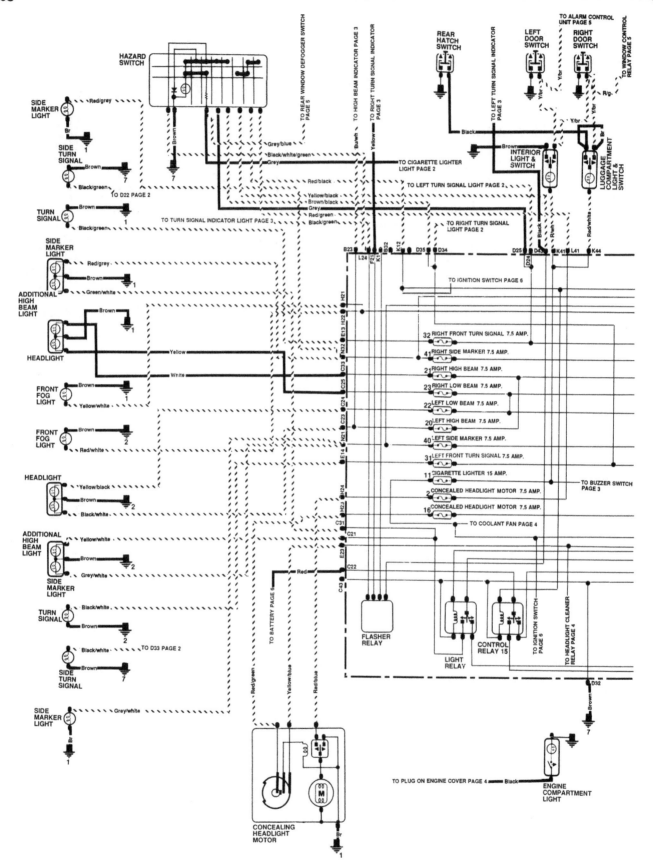

1985 and later models typical: Headlights, side marker lights, turn signals, fog lights, hazard switch and headlight motor
(Page 1 of 7)

0046-H

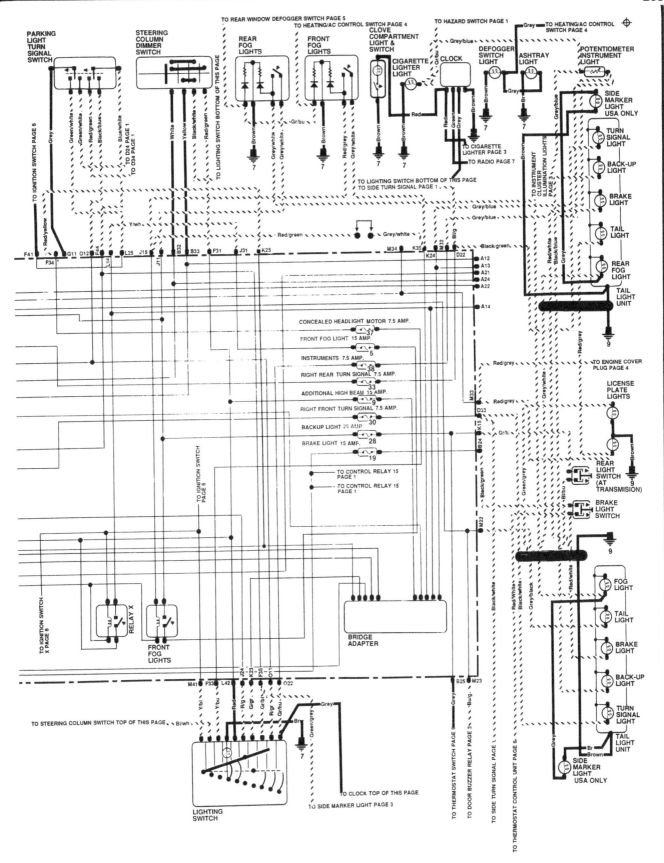

1985 and later models typical: Parking light switch, turn signal switch, dimmer switch, fog lights, interior lights, clock, side marker lights, turn signal lights, back-up lights, brake lights, tail lights and light switch (Page 2 of 7)

0047-H

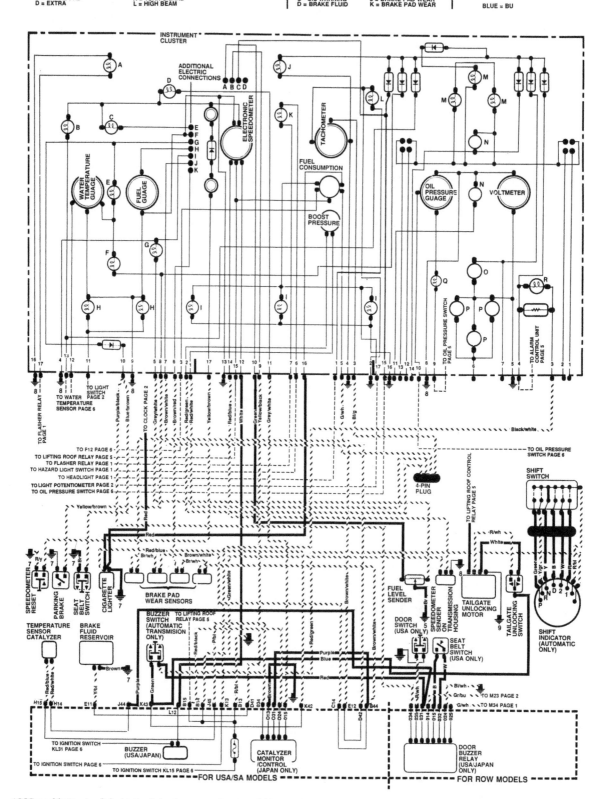

1985 and later models typical: Instrument cluster, shift switch, shift indicator, tailgate switch, tailgate motor, speedometer sensor, fuel level sender, seat belt switch, door buzzer relay, transmission buzzer switch, brake pad wear sensors and cigarette lighter (Page 3 of 7)

0048-H

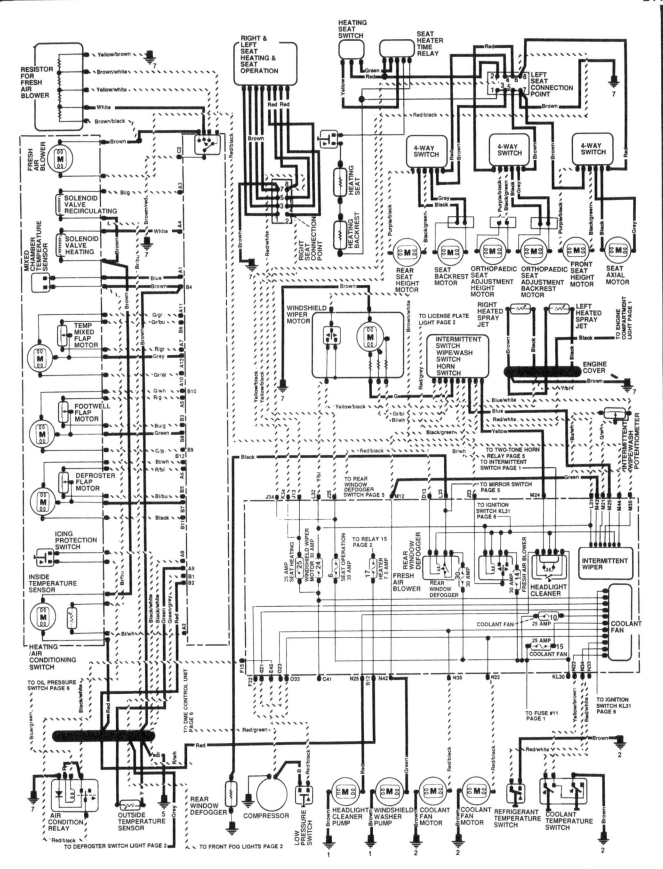

1985 and later models typical: Seat heater, seat adjustment, temperature control/sensors, wiper motor, wiper switch, horn switch, rear window defogger switch, air conditioning, coolant temperature switch and coolant fans (Page 4 of 7)

0049-H

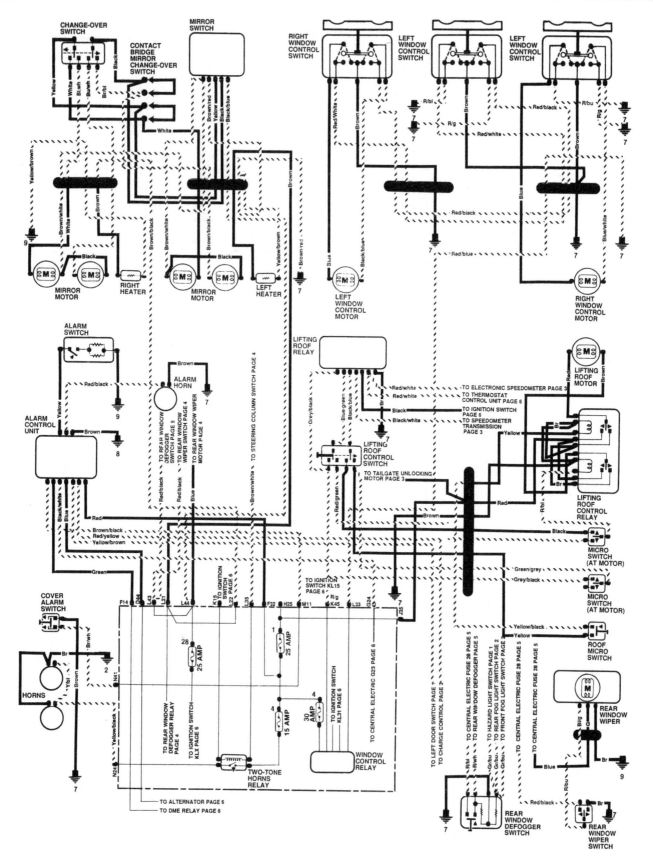

1985 and later models typical: Mirror switch, mirror motor, mirror heater, window controls, alarm system, roof control, horns, rear window wiper and rear window defogger (Page 5 of 7)

0050-H

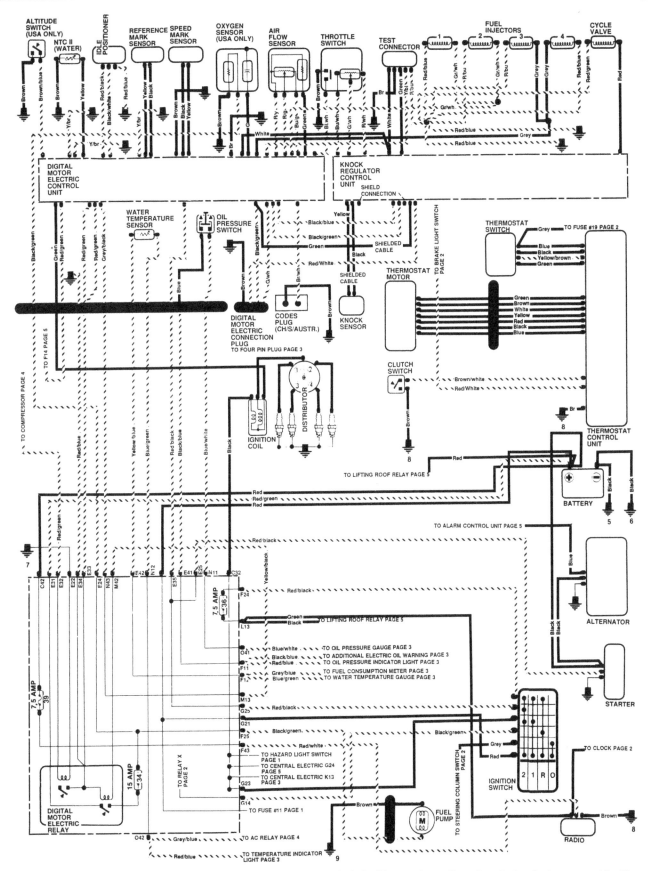

1985 and later models typical: Engine sensors, oil pressure switch, ignition system, alternator, starter, fuel pump and ignition switch (Page 6 of 7)

0051-H

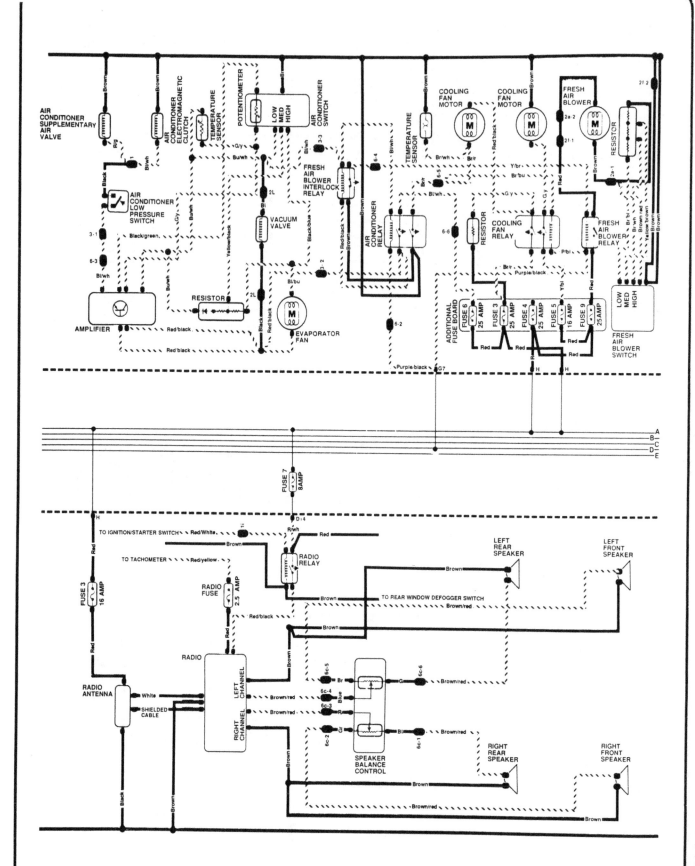

1985 models/1986 typical: Air conditioning, cooling system and radio (Page 7 of 7)

0052-H

Index

Haynes Automotive Manuals

OTE: *If you do not see a listing for your vehicle, please visit* **haynes.com** *for the latest product information and check out our* ***Online Manuals!***

CURA

2020 **Integra** '86 thru '89 **& Legend** '86 thru '90
2021 **Integra** '90 thru '93 **& Legend** '91 thru '95
Integra '94 thru '00 - *see HONDA Civic (42025)*
MDX '01 thru '07 - *see HONDA Pilot (42037)*
2050 **Acura TL** all models '99 thru '08

MC

4020 **Mid-size models** '70 thru '83
4025 **(Renault) Alliance & Encore** '83 thru '87

JDI

5020 **4000** all models '80 thru '87
5025 **5000** all models '77 thru '83
5026 **5000** all models '84 thru '88
Audi A4 '96 thru '01 - *see VW Passat (96023)*
5030 **Audi A4** '02 thru '08

USTIN-HEALEY

Sprite - *see MG Midget (66015)*

MW

8020 **3/5 Series** '82 thru '92
8021 **3-Series** incl. Z3 models '92 thru '98
8022 **3-Series** incl. Z4 models '99 thru '05
8023 **3-Series** '06 thru '14
8025 **320i all 4-cylinder models** '75 thru '83
8050 **1500 thru 2002** except Turbo '59 thru '77

UICK

9010 **Buick Century** '97 thru '05
Century (front-wheel drive) - *see GM (38005)*
9020 **Buick, Oldsmobile & Pontiac Full-size (Front-wheel drive)** '85 thru '05 **Buick** Electra, LeSabre and Park Avenue; **Oldsmobile** Delta 88 Royale, Ninety Eight and Regency; **Pontiac** Bonneville
9025 **Buick, Oldsmobile & Pontiac Full-size (Rear wheel drive)** '70 thru '90 **Buick** Estate, Electra, LeSabre, Limited, **Oldsmobile** Custom Cruiser, Delta 88, Ninety-eight, **Pontiac** Bonneville, Catalina, Grandville, Parisienne
9027 **Buick LaCrosse** '05 thru '13
Enclave - *see GENERAL MOTORS (38001)*
Rainier - *see CHEVROLET (24072)*
Regal - *see GENERAL MOTORS (38010)*
Riviera - *see GENERAL MOTORS (38030, 38031)*
Roadmaster - *see CHEVROLET (24046)*
Skyhawk - *see GENERAL MOTORS (38015)*
Skylark - *see GENERAL MOTORS (38020, 38025)*
Somerset - *see GENERAL MOTORS (38025)*

ADILLAC

21015 **CTS & CTS-V** '03 thru '14
21030 **Cadillac Rear Wheel Drive** '70 thru '93
Cimarron - *see GENERAL MOTORS (38015)*
DeVille - *see GENERAL MOTORS (38031 & 38032)*
Eldorado - *see GENERAL MOTORS (38030)*
Fleetwood - *see GENERAL MOTORS (38031)*
Seville - *see GM (38030, 38031 & 38032)*

HEVROLET

10305 **Chevrolet Engine Overhaul Manual**
24010 **Astro & GMC Safari Mini-vans** '85 thru '05
24013 **Aveo** '04 thru '11
24015 **Camaro V8** all models '70 thru '81
24016 **Camaro** all models '82 thru '92
24017 **Camaro & Firebird** '93 thru '02
Cavalier - *see GENERAL MOTORS (38016)*
Celebrity - *see GENERAL MOTORS (38005)*
24018 **Camaro** '10 thru '15
24020 **Chevelle, Malibu & El Camino** '69 thru '87
Cobalt - *see GENERAL MOTORS (38017)*
24024 **Chevette & Pontiac T1000** '76 thru '87
Citation - *see GENERAL MOTORS (38020)*
24027 **Colorado & GMC Canyon** '04 thru '12
24032 **Corsica & Beretta** all models '87 thru '96
24040 **Corvette** all V8 models '68 thru '82
24041 **Corvette** all models '84 thru '96
24042 **Corvette** all models '97 thru '13
24044 **Cruze** '11 thru '19
24045 **Full-size Sedans** Caprice, Impala, Biscayne, Bel Air & Wagons '69 thru '90
24046 **Impala SS & Caprice and Buick Roadmaster** '91 thru '96
Impala '00 thru '05 - *see LUMINA (24048)*
24047 **Impala & Monte Carlo** all models '06 thru '11
Lumina '90 thru '94 - *see GM (38010)*
24048 **Lumina & Monte Carlo** '95 thru '05
Lumina APV - *see GM (38035)*
24050 **Luv Pick-up** all 2WD & 4WD '72 thru '82
24051 **Malibu** '13 thru '19
24055 **Monte Carlo** all models '70 thru '88
Monte Carlo '95 thru '01 - *see LUMINA (24048)*
24059 **Nova** all V8 models '69 thru '79
24060 **Nova and Geo Prizm** '85 thru '92
24064 **Pick-ups** '67 thru '87 - Chevrolet & GMC
24065 **Pick-ups** '88 thru '98 - Chevrolet & GMC
24066 **Pick-ups** '99 thru '06 - Chevrolet & GMC
24067 **Chevrolet Silverado & GMC Sierra** '07 thru '14
24068 **Chevrolet Silverado & GMC Sierra** '14 thru '19
24070 **S-10 & S-15 Pick-ups** '82 thru '93, **Blazer & Jimmy** '83 thru '94,
24071 **S-10 & Sonoma Pick-ups** '94 thru '04, including **Blazer, Jimmy & Hombre**
24072 **Chevrolet TrailBlazer, GMC Envoy & Oldsmobile Bravada** '02 thru '09
24075 **Sprint** '85 thru '88 **& Geo Metro** '89 thru '01
24080 **Vans - Chevrolet & GMC** '68 thru '96
24081 **Chevrolet Express & GMC Savana** Full-size Vans '96 thru '19

CHRYSLER

10310 **Chrysler Engine Overhaul Manual**
25015 **Chrysler Cirrus, Dodge Stratus, Plymouth Breeze** '95 thru '00
25020 **Full-size Front-Wheel Drive** '88 thru '93
K-Cars - *see DODGE Aries (30008)*
Laser - *see DODGE Daytona (30030)*
25025 **Chrysler LHS, Concorde, New Yorker, Dodge** Intrepid, **Eagle** Vision, '93 thru '97
25026 **Chrysler LHS, Concorde, 300M, Dodge** Intrepid, '98 thru '04
25027 **Chrysler 300** '05 thru '18, **Dodge Charger** '06 thru '18, **Magnum** '05 thru '08 **& Challenger** '08 thru '18
25030 **Chrysler & Plymouth Mid-size** front wheel drive '82 thru '95
Rear-wheel Drive - *see Dodge (30050)*
25035 **PT Cruiser** all models '01 thru '10
25040 **Chrysler Sebring** '95 thru '06, **Dodge Stratus** '01 thru '06 **& Dodge Avenger** '95 thru '00
25041 **Chrysler Sebring** '07 thru '10, **200** '11 thru '17 **Dodge Avenger** '08 thru '14

DATSUN

28005 **200SX** all models '80 thru '83
28012 **240Z, 260Z & 280Z** Coupe '70 thru '78
28014 **280ZX** Coupe & 2+2 '79 thru '83
300ZX - *see NISSAN (72010)*
28018 **510 & PL521 Pick-up** '68 thru '73
28020 **510** all models '78 thru '81
28022 **620 Series Pick-up** all models '73 thru '79
720 Series Pick-up - *see NISSAN (72030)*

DODGE

400 & 600 - *see CHRYSLER (25030)*
30008 **Aries & Plymouth Reliant** '81 thru '89
30010 **Caravan & Plymouth Voyager** '84 thru '95
30011 **Caravan & Plymouth Voyager** '96 thru '02
30012 **Challenger & Plymouth Sapporro** '78 thru '83
30013 **Caravan, Chrysler Voyager & Town & Country** '03 thru '07
30014 **Grand Caravan & Chrysler Town & Country** '08 thru '18
30016 **Colt & Plymouth Champ** '78 thru '87
30020 **Dakota Pick-ups** all models '87 thru '96
30021 **Durango** '98 & '99 **& Dakota** '97 thru '99
30022 **Durango** '00 thru '03 **& Dakota** '00 thru '04
30023 **Durango** '04 thru '09 **& Dakota** '05 thru '11
30025 **Dart, Demon, Plymouth Barracuda, Duster & Valiant** 6-cylinder models '67 thru '76
30030 **Daytona & Chrysler Laser** '84 thru '89
Intrepid - *see CHRYSLER (25025, 25026)*
30034 **Neon** all models '95 thru '99
30035 **Omni & Plymouth Horizon** '78 thru '90
30036 **Dodge & Plymouth Neon** '00 thru '05
30040 **Pick-ups** full-size models '74 thru '93
30042 **Pick-ups** full-size models '94 thru '08
30043 **Pick-ups** full-size models '09 thru '18
30045 **Ram 50/D50 Pick-ups & Raider and Plymouth Arrow Pick-ups** '79 thru '93
30050 **Dodge/Plymouth/Chrysler** RWD '71 thru '89
30055 **Shadow & Plymouth Sundance** '87 thru '94
30060 **Spirit & Plymouth Acclaim** '89 thru '95
30065 **Vans - Dodge & Plymouth** '71 thru '03

EAGLE

Talon - *see MITSUBISHI (68030, 68031)*
Vision - *see CHRYSLER (25025)*

FIAT

34010 **124 Sport Coupe & Spider** '68 thru '78
34025 **X1/9** all models '74 thru '80

FORD

10320 **Ford Engine Overhaul Manual**
10355 **Ford Automatic Transmission Overhaul**
11500 **Mustang '64-1/2 thru '70 Restoration Guide**
36004 **Aerostar Mini-vans** all models '86 thru '97
36006 **Contour & Mercury Mystique** '95 thru '00
36008 **Courier Pick-up** all models '72 thru '82
36012 **Crown Victoria & Mercury Grand Marquis** '88 thru '11
36014 **Edge** '07 thru '19 **& Lincoln MKX** '07 thru '18
36016 **Escort & Mercury Lynx** all models '81 thru '90
36020 **Escort & Mercury Tracer** '91 thru '02
36022 **Escape** '01 thru '17, **Mazda Tribute** '01 thru '11, **& Mercury Mariner** '05 thru '11
36024 **Explorer & Mazda Navajo** '91 thru '01
36025 **Explorer & Mercury Mountaineer** '02 thru '10
36026 **Explorer** '11 thru '17
36028 **Fairmont & Mercury Zephyr** '78 thru '83
36030 **Festiva & Aspire** '88 thru '97
36032 **Fiesta** all models '77 thru '80
36034 **Focus** all models '00 thru '11
36035 **Focus** '12 thru '14
36045 **Fusion** '06 thru '14 **& Mercury Milan** '06 thru '11
36048 **Mustang V8** all models '64-1/2 thru '73
36049 **Mustang II** 4-cylinder, V6 & V8 models '74 thru '78
36050 **Mustang & Mercury Capri** '79 thru '93
36051 **Mustang** all models '94 thru '04
36052 **Mustang** '05 thru '14
36054 **Pick-ups & Bronco** '73 thru '79
36058 **Pick-ups & Bronco** '80 thru '96
36059 **F-150** '97 thru '03, **Expedition** '97 thru '17, **F-250** '97 thru '99, **F-150 Heritage** '04 **& Lincoln Navigator** '98 thru '17
36060 **Super Duty Pick-ups & Excursion** '99 thru '10
36061 **F-150** full-size '04 thru '14
36062 **Pinto & Mercury Bobcat** '75 thru '80
36063 **F-150** full-size '15 thru '17
36064 **Super Duty Pick-ups** '11 thru '16
36066 **Probe** all models '89 thru '92
Probe '93 thru '97 - *see MAZDA 626 (61042)*
36070 **Ranger & Bronco II** gas models '83 thru '92
36071 **Ranger** '93 thru '11 **& Mazda Pick-ups** '94 thru '09
36074 **Taurus & Mercury Sable** '86 thru '95
36075 **Taurus & Mercury Sable** '96 thru '07
36076 **Taurus** '08 thru '14, **Five Hundred** '05 thru '07, **Mercury Montego** '05 thru '07 **& Sable** '08 thru '09
36078 **Tempo & Mercury Topaz** '84 thru '94
36082 **Thunderbird & Mercury Cougar** '83 thru '88
36086 **Thunderbird & Mercury Cougar** '89 thru '97
36090 **Vans** all V8 Econoline models '69 thru '91
36094 **Vans** full size '92 thru '14
36097 **Windstar** '95 thru '03, **Freestar & Mercury Monterey Mini-van** '04 thru '07

GENERAL MOTORS

10360 **GM Automatic Transmission Overhaul**
38001 **GMC Acadia** '07 thru '16, **Buick Enclave** '08 thru '17, **Saturn Outlook** '07 thru '10 **& Chevrolet Traverse** '09 thru '17
38005 **Buick Century, Chevrolet Celebrity, Oldsmobile Cutlass Ciera & Pontiac 6000** all models '82 thru '96
38010 **Buick Regal** '88 thru '04, **Chevrolet Lumina** '88 thru '04, **Oldsmobile Cutlass Supreme** '88 thru '97 **& Pontiac Grand Prix** '88 thru '07
38015 **Buick Skyhawk, Cadillac Cimarron, Chevrolet Cavalier, Oldsmobile Firenza, Pontiac J-2000 & Sunbird** '82 thru '94
38016 **Chevrolet Cavalier & Pontiac Sunfire** '95 thru '05
38017 **Chevrolet Cobalt** '05 thru '10, **HHR** '06 thru '11, **Pontiac G5** '07 thru '09, **Pursuit** '05 thru '06 **& Saturn ION** '03 thru '07
38020 **Buick Skylark, Chevrolet Citation, Oldsmobile Omega, Pontiac Phoenix** '80 thru '85
38025 **Buick Skylark** '86 thru '98, **Somerset** '85 thru '87, **Oldsmobile Achieva** '92 thru '98, **Calais** '85 thru '91, **& Pontiac Grand Am** all models '85 thru '98
38026 **Chevrolet Malibu** '97 thru '03, **Classic** '04 thru '05, **Oldsmobile Alero** '99 thru '03, **Cutlass** '97 thru '00, **& Pontiac Grand Am** '99 thru '03
38027 **Chevrolet Malibu** '04 thru '12, **Pontiac G6** '05 thru '10 **& Saturn Aura** '07 thru '10
38030 **Cadillac Eldorado, Seville, Oldsmobile Toronado & Buick Riviera** '71 thru '85
38031 **Cadillac Eldorado, Seville, DeVille, Fleetwood, Oldsmobile Toronado & Buick Riviera** '86 thru '93
38032 **Cadillac DeVille** '94 thru '05, **Seville** '92 thru '04 **& Cadillac DTS** '06 thru '10
38035 **Chevrolet Lumina APV, Oldsmobile Silhouette & Pontiac Trans Sport** all models '90 thru '96
38036 **Chevrolet Venture** '97 thru '05, **Oldsmobile Silhouette** '97 thru '04, **Pontiac Trans Sport** '97 thru '98 **& Montana** '99 thru '05
38040 **Chevrolet Equinox** '05 thru '17, **GMC Terrain** '10 thru '17 **& Pontiac Torrent** '06 thru '09

GEO

Metro - *see CHEVROLET Sprint (24075)*
Prizm - *'85 thru '92 see CHEVY (24060), '93 thru '02 see TOYOTA Corolla (92036)*
40030 **Storm** all models '90 thru '93
Tracker - *see SUZUKI Samurai (90010)*

(Continued on other side)

Haynes Automotive Manuals (continued)

NOTE: If you do not see a listing for your vehicle, please visit **haynes.com** for the latest product information and check out our **Online Manuals**

GMC
Acadia - see GENERAL MOTORS (38001)
Pick-ups - see CHEVROLET (24027, 24068)
Vans - see CHEVROLET (24081)

HONDA
42010 **Accord CVCC** all models '76 thru '83
42011 **Accord** all models '84 thru '89
42012 **Accord** all models '90 thru '93
42013 **Accord** all models '94 thru '97
42014 **Accord** all models '98 thru '02
42015 **Accord** '03 thru '12 **& Crosstour** '10 thru '14
42016 **Accord** '13 thru '17
42020 **Civic 1200** all models '73 thru '79
42021 **Civic 1300 & 1500 CVCC** '80 thru '83
42022 **Civic 1500 CVCC** all models '75 thru '79
42023 **Civic** all models '84 thru '91
42024 **Civic & del Sol** '92 thru '95
42025 **Civic** '96 thru '00, **CR-V** '97 thru '01 **& Acura Integra** '94 thru '00
42026 **Civic** '01 thru '11 **& CR-V** '02 thru '11
42027 **Civic** '12 thru '15 **& CR-V** '12 thru '16
42030 **Fit** '07 thru '13
42035 **Odyssey** all models '99 thru '10
Passport - see ISUZU Rodeo (47017)
42037 **Honda Pilot** '03 thru '08, **Ridgeline** '06 thru '14 **& Acura MDX** '01 thru '07
42040 **Prelude CVCC** all models '79 thru '89

HYUNDAI
43010 **Elantra** all models '96 thru '19
43015 **Excel & Accent** all models '86 thru '13
43050 **Santa Fe** all models '01 thru '12
43055 **Sonata** all models '99 thru '14

INFINITI
G35 '03 thru '08 - see NISSAN 350Z (72011)

ISUZU
Hombre - see CHEVROLET S-10 (24071)
47017 **Rodeo** '91 thru '02, **Amigo** '89 thru '94 & '98 thru '02 **& Honda Passport** '95 thru '02
47020 **Trooper** '84 thru '91 **& Pick-up** '81 thru '93

JAGUAR
49010 **XJ6** all 6-cylinder models '68 thru '86
49011 **XJ6** all models '88 thru '94
49015 **XJ12 & XJS** all 12-cylinder models '72 thru '85

JEEP
50010 **Cherokee, Comanche & Wagoneer Limited** all models '84 thru '01
50011 **Cherokee** '14 thru '19
50020 **CJ** all models '49 thru '86
50025 **Grand Cherokee** all models '93 thru '04
50026 **Grand Cherokee** '05 thru '19 **& Dodge Durango** '11 thru '19
50029 **Grand Wagoneer & Pick-up** '72 thru '91 Grand Wagoneer '84 thru '91, Cherokee & Wagoneer '72 thru '83, Pick-up '72 thru '88
50030 **Wrangler** all models '87 thru '17
50035 **Liberty** '02 thru '12 **& Dodge Nitro** '07 thru '11
50050 **Patriot & Compass** '07 thru '17

KIA
54050 **Optima** '01 thru '10
54060 **Sedona** '02 thru '14
54070 **Sephia** '94 thru '01, **Spectra** '00 thru '09, **Sportage** '05 thru '20
54077 **Sorento** '03 thru '13

LEXUS
ES 300/330 - see TOYOTA Camry (92007, 92008)
ES 350 - see TOYOTA Camry (92009)
RX 300/330/350 - see TOYOTA Highlander (92095)

LINCOLN
MKX - see FORD (36014)
Navigator - see FORD Pick-up (36059)
59010 **Rear-Wheel Drive Continental** '70 thru '87, **Mark Series** '70 thru '92 **& Town Car** '81 thru '10

MAZDA
61010 **GLC** (rear-wheel drive) '77 thru '83
61011 **GLC** (front-wheel drive) '81 thru '85
61012 **Mazda3** '04 thru '11
61015 **323 & Protegé** '90 thru '03
61016 **MX-5 Miata** '90 thru '14
61020 **MPV** all models '89 thru '98
Navajo - see Ford Explorer (36024)
61030 **Pick-ups** '72 thru '93
Pick-ups '94 thru '09 - see Ford Ranger (36071)
61035 **RX-7** all models '79 thru '85
61036 **RX-7** all models '86 thru '91
61040 **626** (rear-wheel drive) all models '79 thru '82
61041 **626 & MX-6** (front-wheel drive) '83 thru '92
61042 **626** '93 thru '01 **& MX-6/Ford Probe** '93 thru '02
61043 **Mazda6** '03 thru '13

MERCEDES-BENZ
63012 **123 Series Diesel** '76 thru '85
63015 **190 Series** 4-cylinder gas models '84 thru '88
63020 **230/250/280** 6-cylinder SOHC models '68 thru '72
63025 **280 123 Series** gas models '77 thru '81
63030 **350 & 450** all models '71 thru '80
63040 **C-Class:** C230/C240/C280/C320/C350 '01 thru '07

MERCURY
64200 **Villager & Nissan Quest** '93 thru '01
All other titles, see FORD Listing.

MG
66010 **MGB** Roadster & GT Coupe '62 thru '80
66015 **MG Midget, Austin Healey Sprite** '58 thru '80

MINI
67020 **Mini** '02 thru '13

MITSUBISHI
68020 **Cordia, Tredia, Galant, Precis & Mirage** '83 thru '93
68030 **Eclipse, Eagle Talon & Plymouth Laser** '90 thru '94
68031 **Eclipse** '95 thru '05 **& Eagle Talon** '95 thru '98
68035 **Galant** '94 thru '12
68040 **Pick-up** '83 thru '96 **& Montero** '83 thru '93

NISSAN
72010 **300ZX** all models including Turbo '84 thru '89
72011 **350Z & Infiniti G35** all models '03 thru '08
72015 **Altima** all models '93 thru '06
72016 **Altima** '07 thru '12
72020 **Maxima** all models '85 thru '92
72021 **Maxima** all models '93 thru '08
72025 **Murano** '03 thru '14
72030 **Pick-ups** '80 thru '97 **& Pathfinder** '87 thru '95
72031 **Frontier** '98 thru '04, **Xterra** '00 thru '04, **& Pathfinder** '96 thru '04
72032 **Frontier & Xterra** '05 thru '14
72037 **Pathfinder** '05 thru '14
72040 **Pulsar** all models '83 thru '86
72042 **Roque** all models '08 thru '20
72050 **Sentra** all models '82 thru '94
72051 **Sentra & 200SX** all models '95 thru '06
72060 **Stanza** all models '82 thru '90
72070 **Titan pick-ups** '04 thru '10, **Armada** '05 thru '10 **& Pathfinder Armada** '04
72080 **Versa** all models '07 thru '19

OLDSMOBILE
73015 **Cutlass** V6 & V8 gas models '74 thru '88
For other OLDSMOBILE titles, see BUICK, CHEVROLET or GENERAL MOTORS listings.

PLYMOUTH
For PLYMOUTH titles, see DODGE listing.

PONTIAC
79008 **Fiero** all models '84 thru '88
79018 **Firebird** V8 models except Turbo '70 thru '81
79019 **Firebird** all models '82 thru '92
79025 **G6** all models '05 thru '09
79040 **Mid-size Rear-wheel Drive** '70 thru '87
Vibe '03 thru '10 - see TOYOTA Corolla (92037)
For other PONTIAC titles, see BUICK, CHEVROLET or GENERAL MOTORS listings.

PORSCHE
80020 **911** Coupe & Targa models '65 thru '89
80025 **914** all 4-cylinder models '69 thru '76
80030 **924** all models including Turbo '76 thru '82
80035 **944** all models including Turbo '83 thru '89

RENAULT
Alliance & Encore - see AMC (14025)

SAAB
84010 **900** all models including Turbo '79 thru '88

SATURN
87010 **Saturn** all S-series models '91 thru '02
Saturn Ion '03 thru '07- see GM (38017)
Saturn Outlook - see GM (38001)
87020 **Saturn L-series** all models '00 thru '04
87040 **Saturn VUE** '02 thru '09

SUBARU
89002 **1100, 1300, 1400 & 1600** '71 thru '79
89003 **1600 & 1800** 2WD & 4WD '80 thru '94
89080 **Impreza** '02 thru '11, **WRX** '02 thru '14, **& WRX STI** '04 thru '14
89100 **Legacy** all models '90 thru '99
89101 **Legacy & Forester** '00 thru '09
89102 **Legacy** '10 thru '16 **& Forester** '12 thru '16

SUZUKI
90010 **Samurai/Sidekick & Geo Tracker** '86 thru '01

TOYOTA
92005 **Camry** all models '83 thru '91
92006 **Camry** '92 thru '96 **& Avalon** '95 thru '96
92007 **Camry, Avalon, Solara, Lexus ES 300** '97 thru '01
92008 **Camry, Avalon, Lexus ES 300/330** '02 thru '06 **& Solara** '02 thru '08
92009 **Camry, Avalon & Lexus ES 350** '07 thru '17
92015 **Celica Rear-wheel Drive** '71 thru '85
92020 **Celica Front-wheel Drive** '86 thru '99
92025 **Celica Supra** all models '79 thru '92
92030 **Corolla** all models '75 thru '79
92032 **Corolla** all rear-wheel drive models '80 thru '87
92035 **Corolla** all front-wheel drive models '84 thru '92
92036 **Corolla & Geo/Chevrolet Prizm** '93 thru '02
92037 **Corolla** '03 thru '19, **Matrix** '03 thru '14, **& Pontiac Vibe** '03 thru '10
92040 **Corolla Tercel** all models '80 thru '82
92045 **Corona** all models '74 thru '82
92050 **Cressida** all models '78 thru '82
92055 **Land Cruiser FJ40, 43, 45, 55** '68 thru '82
92056 **Land Cruiser FJ60, 62, 80, FZJ80** '80 thru '96
92060 **Matrix** '03 thru '11 **& Pontiac Vibe** '03 thru '10
92065 **MR2** all models '85 thru '87
92070 **Pick-up** all models '69 thru '78
92075 **Pick-up** all models '79 thru '95
92076 **Tacoma** '95 thru '04, **4Runner** '96 thru '02 **& T100** '93 thru '08
92077 **Tacoma** all models '05 thru '18
92078 **Tundra** '00 thru '06 **& Sequoia** '01 thru '07
92079 **4Runner** all models '03 thru '09
92080 **Previa** all models '91 thru '95
92081 **Prius** all models '01 thru '12
92082 **RAV4** all models '96 thru '12
92085 **Tercel** all models '87 thru '94
92090 **Sienna** all models '98 thru '10
92095 **Highlander** '01 thru '19 **& Lexus RX330/330/350** '99 thru '19
92179 **Tundra** '07 thru '19 **& Sequoia** '08 thru '19

TRIUMPH
94007 **Spitfire** all models '62 thru '81
94010 **TR7** all models '75 thru '81

VW
96008 **Beetle & Karmann Ghia** '54 thru '79
96009 **New Beetle** '98 thru '10
96016 **Rabbit, Jetta, Scirocco & Pick-up** gas models '75 thru '92 & Convertible '80 thru '9
96017 **Golf, GTI & Jetta** '93 thru '98, **Cabrio** '95 thru '02
96018 **Golf, GTI, Jetta** '99 thru '05
96019 **Jetta, Rabbit, GLI, GTI & Golf** '05 thru '11
96020 **Rabbit, Jetta & Pick-up** diesel '77 thru '84
96021 **Jetta** '11 thru '18 **& Golf** '15 thru '19
96023 **Passat** '98 thru '05 **& Audi A4** '96 thru '01
96030 **Transporter 1600** all models '68 thru '79
96035 **Transporter 1700, 1800 & 2000** '72 thru '79
96040 **Type 3 1500 & 1600** all models '63 thru '73
96045 **Vanagon Air-Cooled** all models '80 thru '83

VOLVO
97010 **120, 130 Series & 1800 Sports** '61 thru '73
97015 **140 Series** all models '66 thru '74
97020 **240 Series** all models '76 thru '93
97040 **740 & 760 Series** all models '82 thru '88
97050 **850 Series** all models '93 thru '97

TECHBOOK MANUALS
10205 **Automotive Computer Codes**
10206 **OBD-II & Electronic Engine Management**
10210 **Automotive Emissions Control Manual**
10215 **Fuel Injection Manual** '78 thru '85
10225 **Holley Carburetor Manual**
10230 **Rochester Carburetor Manual**
10305 **Chevrolet Engine Overhaul Manual**
10320 **Ford Engine Overhaul Manual**
10330 **GM and Ford Diesel Engine Repair Manual**
10331 **Duramax Diesel Engines** '01 thru '19
10332 **Cummins Diesel Engine Performance Manual**
10333 **GM, Ford & Chrysler Engine Performance Manual**
10334 **GM Engine Performance Manual**
10340 **Small Engine Repair Manual**, 5 HP & Less
10341 **Small Engine Repair Manual**, 5.5 HP 20 HP
10345 **Suspension, Steering & Driveline Manual**
10355 **Ford Automatic Transmission Overhaul**
10360 **GM Automatic Transmission Overhaul**
10405 **Automotive Body Repair & Painting**
10410 **Automotive Brake Manual**
10411 **Automotive Anti-lock Brake (ABS) Systems**
10420 **Automotive Electrical Manual**
10425 **Automotive Heating & Air Conditioning**
10435 **Automotive Tools Manual**
10445 **Welding Manual**
10450 **ATV Basics**

Over a 100 Haynes motorcycle manuals also available

10/22